Cai Yuanpei Project 2012 Number 28007 UF

Medical Law and Bioethics Comparison between France and China in The Application to Genetics, Biotechnologies and Public Health

THE BOOK OF CAI YUANPEI PROJECT

BIOTECHNOLOGY, MEDICINE AND LAW

生物科技、医学与法律

Qi Yanping Anne-Marie Duguet / Chief Editor

齐延平　安 · 玛丽 · 杜盖 / 主编

Man Hongjie Chen Zihan / Executive Editior

满洪杰　陈姿含 / 执行主编

中国政法大学出版社

2018 · 北京

图书在版编目（CIP）数据

生物科技、医学与法律 = Biotechnology,Medicineand Law：英文/齐延平，(法）安·玛丽·杜盖主编.—北京：中国政法大学出版社，2018.8（2022.9重印）

ISBN 978-7-5620-8420-4

Ⅰ.①生… Ⅱ.①齐… ②安… Ⅲ.①生物工程—研究—英文 Ⅳ.①Q81-49

中国版本图书馆CIP数据核字(2018)第201750号

出版者	中国政法大学出版社
地　址	北京市海淀区西土城路 25 号
邮寄地址	北京 100088 信箱 8034 分箱　邮编 100088
网　址	http://www.cuplpress.com (网络实名：中国政法大学出版社)
电　话	010-58908524(编辑部) 58908334(邮购部)
承　印	北京九州迅驰传媒文化有限公司
开　本	720mm×960mm　1/16
印　张	19.25
字　数	310 千字
版　次	2018 年 8 月第 1 版
印　次	2022 年 9 月第 2 次印刷
定　价	78.00 元

Preface

It is our great pleasure to introduce this book, the fruit of the cooperation between the Law School of Shandong University and the INSERM Unit 1027 of Paul Sabatier University.

This cooperation was organized in the context of the Hubert Curien partnership with China, Program Cai Yuanpei, coordinated by Campus France. Both teams had researchers interested in health law and bioethics. The originality of the program lies in the exchanges of PhD students who have met regularly both in France and China, to discuss their research and to write comparative articles.

Two co – supervised theses were conducted, directed by Dean QI Yanping and Dr. Anne – Marie Duguet. PhD students Zihan Chen and Meng Wen made considerable efforts during their stay in France to adapt to the lifestyle and especially to meet the French requirements and academic control of their research. The theses defended in China were issued under the double seal of Paul Sabatier University and Law School of Shandong University.

On the French side, French – speaking PhD students were selected: Mou Li and Chuanjuan Zhuang. Their theses were co – supervised by Dr Anne – Marie Duguet, Pr Jacques Larrieu and Pr Alexandra Mendoza – Caminade. They traveled to China on several occasions to deepen their research, especially Chuanjuan Zuhang whose thesis focuses on traditional Chinese medicine.

Our project has also allowed visits of Chinese senior researchers in France and vice versa. A workshop was organized by Prof. Hongjie Man at the Law School of Shandong University on the theme "Biotechnology, medicine and law" and this book brings together the main works presented. Most authors are lawyers but as well doctors, geneticists, epidemiologists, public health specialists. The multidisciplinarity of our teams has enriched the quality of the work. Regarding the authors of the French team, their contributions are divided into 3 parts of the book.

The first part deals with the genetic information and its protection which are ensured in Europe by recommendations and a very strict legal framework of the Council of Europe. Thus free access genetic testing that is offered in many countries around the world is not possible in France.

The second part deals with the ethical aspects of biotechnology. All countries are concerned with questions of gene patentability, stem cell use and embryo research. The fundamental principles of the dignity of the human being have no boundaries and the protection of individual rights prevails over the interest of science.

The third part presents the challenges of pharmaceutical law, biotechnology and new therapeutics, but also those of intellectual property law. Finally, China remains very attractive for foreigners as evidenced by the last two articles one medical tourism and the other on the protection of traditional Chinese medicine.

On the Chinese side, the Director of PhD students, Pr. Qi Yanping, Director of the Centre for Human Rights Study at Shandong University, contributes to researching the rights systems and has a deep thinking on the issues while the traditional legal system facing the challenges in big data era. Pr. Wang Kang, working in Shanghai University of Political Science and Law, PhD in Law is the scholar studying the biomedicine and its legal issues. The writing of his doctoral dissertation opens a new field of vision in related fields. More and more young scholars such as Li Yan and Jin Genlin are also involved in the research of this field, and contribute their own thinking, dates and unique analytical methods.

This cooperation was a unique experience thanks to the participation of French –

speaking Chinese colleagues. In our group, Prof. Wu Tao, who studied and defended his PhD in France, was a great help for the construction of the project and the mutual understanding during the symposiums and the workshops. The various meetings in France and China have created a France – Europe – China network of researchers and PhD students in health law who wish to continue their collaborations in the future.

Our deep gratitude goes to the Embassy of France in China for its support and especially to Mrs Christine Da Luz and Mrs Chin By Ang who have encouraged us throughout this project.

Pr Qi Yanping
Chair Professor of Law and Director of
The Centre for Human Rights Study
Shandong University

Dr Anne – Marie Dugue
MD PhD Emeritus Senior lecturer
UMR/INSERM Unit 1027
Paul Sabatier University

Contents

Part One The Protection of Genetic Data from the Comparative Perspective

PROTECTION OF GENETIC INFORMATION IN EUROPE AND IN FRANCE

Anne – Marie Duguet[1], Emmanuelle Rial[2], Anne Cambon – Thomsen[3][4]

Genetic characteristics are elements of the personality and genetic analysis is restricted in some countries to the scientific or medical purpose and performed on medical prescription. Nevertheless, many people believe that genetics will predict the evolution of their health. As private genetic tests are now accessible through the internet, there is a real demand for open genetic testing and for free access to genetic information.

It is important for individuals to understand the consequences of a genetic test and the information collected, they will need support and help in knowing what to do with the information, once they decide they want it. On the one hand, the genetic information can help to make decisions about possible prevention or therapy, but on the other hand this information can also impose an enormous psychological burden on patients and their families, if there is no known way of preventing or treating the condition. Aggressively marketed genetic tests, for which evidence of their usefulness is limited or absent, may heighten anxieties about health or lead to inappropriate re-

[1] MD, PhD Senior lecturer Faculty of Medicine

[2] Lawyer PhD Research assistant

[3] CNRS Research Director

[4] UMR U 1027, Inserm, Université de Toulouse – Université Paul Sabatier – Toulouse III, Epidémiologie et analyses en santé publique : risques, maladies chroniques et handicap
Département d'épidémiologie et de santé publique
Faculté de médecine. 37 allées Jules Guesde F – 31073 Toulouse Cedex 7
aduguet@ club – internet. fr

quests for further medical tests or treatment.

In the UK, the Human Genetics Commission HGC[1] pointed out that concerns have been raised regarding the tests' efficacy, utility and implications for individuals and their families. Tests that claim to predict the onset of disease or indicate a heightened risk of serious conditions need to be evidenced that they accurately and reliably predict what they are advertised as predicting. They also need to be provided in the context of proper consultation where their implications can be discussed and managed[2].

Genetic tests in free access are new concerns but genetic analysis in general has been debated by many international organizations who published guidelines, among them the Council of Europe, in the Oviedo convention. (1997). The article 11 says that no discrimination should be made according to the genetic characteristics, and the article 12 restricts the genetic diagnosis to scientific or medical purposes with an appropriate genetic counseling.

The Council of Europe is very protective for the right of the persons, but the provisions of the Convention are not respected in countries in Europe since open practices are possible such as private genetic tests accessible to the public, without medical prescription.

After a definition of genetic testing, genetic information and genetic data, we will present through the Oviedo convention and its protocol the position of the Council of Europe (A).

As the situation differs within countries, even in the EU, we present the opinion of international organizations allowing national provisions of the law. In this respect the French legislation is very specific and more protective for genetic characteristics

---

[1] The HGC is the UK Government's advisory body on new developments in human genetics and how they impact on individual lives. http://www.hgc.gov.uk.

[2] On 4 August, 2010, The HGC published a "Common Framework of Principles" for direct-to-consumer genetic testing service.

(B).

A – Genetic information, ethical questions and the Oviedo Convention

Genetic information is all the information collected through the analysis of genetic characteristics. A distinction should be made on the purpose of genctic analysis, either medical or scientific.

– For medical purpose, the analysis of genetic characteristics can be a genetic diagnosis which is the identification of one or several gene (s) responsible for a disease, or a genetic predictive diagnosis which is the detection of a genetic predisposition or susceptibility to a disease. Genetic analysis can diagnose the genetic disease on a patient presenting symptoms as well as the transmission of the disease to the child to be born.

The subject who does not present any symptoms can be a carrier of the gene responsible for a disease, or have a predisposition or susceptibility for a genetic disease, and then the analysis of genetic characteristics is a predictive diagnosis.

– The scientific purpose can be research protocols on genetics or genetic screening.

A research protocol on genetics is conducted on biological samples for identification of new anomalies on the genes or of biomarkers of the disease. The research may be focused on population and groups, or be performed with genetic biobanks. The research implies sometimes the secondary use of biological samples and genetic data collected during the care to build large biobanks or genetic data banks.

A genetic screening is the systematic search in a population for individuals with a particular genetic disease or genotypes, and provides collective information.

The research's purpose is not to give a diagnosis of a disease to the person involved in research, but to increase the knowledge of the disease by the identification of predispositions or new biological markers. No individual information is supposed to be given to the subject. This is a concern for researchers working on new technologies because it is difficult to determine the limit between research and clinic and the information of the individual results becomes complex[1]. What to say with the incidental findings? What to do of uncertain results?

The American college of human genetics[2] has established guidelines for testing and reporting ultra rare disorders and calls for caution and for restriction of results sharing. Faced with uncertainty, the provider may be obliged to avoid the possibility of harm rather than to provide unclear benefits. There may be rebuttable presumption to defer testing unless the risk/benefit ratio is favorable.

I – Ethical issues in genetic testing and information

The first main issue is that the public does not make the difference between the test conducted for medical purposes on medical prescription and the private or free access genetic tests performed without any guidelines. Public should be aware of the difference between diagnosis and predictive tests, and of the level of quality insurance of the tests.

Recognizing the importance of ethical legal social issues in human genomics, The Human Genetics programme (HGN) identified genetic testing and screening as a priority area with the following issues[3]: information and genetic counseling, confidentiality, stigmatization and discrimination, and gives the following comments:

> It is therefore very important that genetic screening and testing are accompanied by education and counseling. Genetic tests give an assessment of an individual's inherent risk for disease and disability. This predictive power makes genetic testing particularly liable for misuse. Employers and insurance compa-

[1] A. Soulier, S. Julia, A. Cambon – Thomsen, "A Review of Ethical Questions Raised by the Transfer into Clinics of Massive Parallel Sequencing Technologies", in Droits des patients, mobilité et accès aux soins. Book coordinated by AM Duguet Les Etudes Hospitalières 2011.

[2] http://www.acmg.net/Pages/ACMG_ Activities/stds – 2002/URD.htm.

[3] WHO, http://www.who.int/genomics.

nies have been known to deny individuals essential health care or employment based on knowledge of genetic disposition. This type of discrimination can be socially debilitating and has severe socio – economic consequences. It is important, therefore, to ensure the confidentiality of test results, and to establish legislation permitting only selective access to this information.

Genetic information can have important implications not only for the one who is tested, but also for his or her relatives. Respecting a patient's confidentiality by not disclosing the results of a genetic test to third parties can therefore conflict with the well – being of family members, who could benefit from this knowledge.

Knowledge of risk of disease may be used by health insurance providers and employers to deny individuals employment, benefits and allowances and medical coverage or health insurance. This is especially worrisome in communities that rely heavily on private insurance systems as a source of funding for necessary medical treatments

1) Which benefit for the person to know the results of genetic testing?

Genetic susceptibility testing for multi – factorial diseases usually has limited clinical utility for diseases that are currently untreatable and for which no preventive measures are available to prevent symptoms or to delay the disease onset.

At the opposite, in the context of predictive testing for monogenic late onset diseases, a special attention has been paid[1] to the psychological and ethical dimension of predictive testing for Huntington Disease in adults as well as to prenatal diagnosis during pregnancy and pre – implantation genetic diagnosis in embryos (PGD)[2].

[1] See Gerry Evers – Kiebooms, Department of Human Genetics, Leuven, Belgium "Predictive genetic tests in practice: a psychological and ethical perspective" presentation to the Academic Session of the European Summer School on health law and bioethics. Toulouse July 2011.

[2] The International Huntington Association has formulated guidelines for PGD. www. huntington – assoc. com

Performing predictive testing for hereditary cancers[1] is justified by the availability of preventive measures (regular surveillance aimed at early diagnosis or preventive surgery).

2) Who delivers the information ? The promotion of genetic counseling

For medical purposes, clinical genetic services are now frequently provided by multidisciplinary teams. In many European countries, nonmedical genetic counselors are working as part of the clinical team, providing both information and support for individuals and families at risk of or affected by a genetic condition[2].

Genetic counseling is a vital part of the field of medical genetics. Genetic counselors are trained in the fields of genetics and psychosocial counseling, and act as advocates for families affected by genetic disorders, helping them to understand the concepts of heredity and assisting them in planning for treatment of affected individuals as well as providing options for future offspring.

3) Which access for third parties to the genetic information? Discrimination and insurance companies

Insurers wish to share genetic health information with clients to enable underwriters to make an accurate assessment of the risk[3]. So Insurance companies would like to have direct access to the results of the test or be allowed to require the test, or ask subjects who have performed the test to provide the results before the signature of the insurance contract.

Genetic technology increases opportunity for individuals to obtain additional risk estimates about their susceptibilities to disease. Insurers argue that they need to have access to any information that predicts disease risk because the amount that policyholders pay for insurance coverage is determined by assessing their level of risk.

---

[1] Hereditary breast - and ovarian cancer caused by a mutation in the BRCA1 - or BRCA2 gene.

[2] See Heater Skirton Genetic counseling profession in Europe, Published on line September 2009. Full article on Wiley Online library.

[3] Yvonne Bombard, Trudo Lemmens, "Insurance and Genetic Information" Published online: April 2010 DOI: 10.1002/9780470015902.a0005203.pub2 Full Article on Wiley Online Library.

However, people are reluctant to share genetic test results with insurers due to the potential risk of insurance discrimination because population groups will be denied insurance because of their genetic profiles. The situation is unclear in many countries with insurance companies, especially with health insurance companies. In France, following the opinion of the French Ethics committee, the insurance companies cannot ask for this information and now the law forbids any question on this purpose. (see below)

II – The Oviedo convention and its additional protocol on genetic tests

The convention on human rights and biomedicine (4 April 1997)[1] sets up general principles for genetics in 4 articles and has been completed by the additional protocol on genetic tests.

1) The Oviedo convention

The article 11 of the Convention is related to the non – discrimination principle: "Any form of discrimination against a person on grounds of his or her genetic heritage is prohibited" and the article 12 defines the predictive genetic tests, and require an appropriate genetic counseling: "Tests which are predictive of genetic disease or which serve either to identify the subject as a carrier of a gene responsible for a disease or to detect a genetic predisposition or susceptibility to a disease may be performed only for health purposes or for scientific research linked with health purposes, and subject to appropriate genetic counseling."

The article 13 limits the intervention on human genome: "Any intervention seeking to modify the human genome may only be undertaken for preventive, diagnostic or therapeutic purposes and only if it aims is not to introduce any modification in the genome of any descendants."

And the article 14 forbids the selection of sex: "The use of techniques of medi-

[1] Convention for the protection of human rights and dignity of the human being with regard to the application of biology and medicine, http://conventions.coe.int.

cally assisted procreation shall not be allowed for the purpose of choosing a future child's sex, except where serious hereditary sex - related diseases is to be avoided. "

2) The additional protocol on genetics tests

10 years later, as new developments in genetics occurred especially in genetics testing, the additional protocol has been adopted on 27 November, 2008.[1]

In the preamble, the protocol acknowledges the benefit of genetics, in particular genetic testing, in the field of health, but recognizes that concerns exist regarding possible improper use of genetic testing, in particular of the information generated thereby.

The scope of the protocol is limited to tests which are carried out for health purposes, involving analysis of biological samples of human origin and aiming specifically to identify the genetic characteristics of a person which are inherited or acquired during early prenatal development. The protocol is not applicable to tests performed for research purposes.

General provisions are recalled such as the welfare of human being (art 3) and the prohibition of discrimination or stigmatization (art 4).

The protocol reminds the necessary measures to insure appropriate quality of genetic services (art 5), with criteria of scientific utility, the quality insurance of the services and the qualification of persons providing the tests.

The offer of the test should have a clinical utility (art 6) and should be performed under medical personalized supervision (art 7).

Regarding the information (art 8), the protocol requires in any case a documented information especially on the implication of the results before consent. Moreover, a written consent is necessary for predictive tests. In addition the protocol imposes a genetic counseling according to the implications of the results of the predictive tests and their significance for the person or the members of his or her family,

---

[1] http://conventions.coe.int/treaty/EN/treaties/Html/203.htm.

including possible implications concerning procreation choices. The person concerned may freely withdraw consent at any time. (art 9)

As for the results, the subject is entitled to know any information[1], the conclusions must be presented in comprehensive form (art 16) and the wish not to be informed must be respected.

The protocol describes the conditions of consent for minors and person unable to consent in articles 10, 11, 12. Then, a legal representative will give the consent. The opinion of the minor or of the incompetent person shall be taken into consideration as an increasingly determining factor in proportion to his or her age and degree of maturity or understanding. No test can be performed if the person expressly opposed such test.

The confidentiality of the results is imperative. Everyone has the right to be respected for his or her private life, in particular to protection of his or her personal data derived from a genetic test (art 16).

Through these two instruments, the Council of Europe organized a very good protection of the rights of individuals. Nevertheless, the development of new technologies and their commercial application enlarge the debate and the major international organizations provided declarations.

B – Opinions of European and international organizations and the French particularities

I – International guidelines

For international organizations the basic principles are similar as those of the Oviedo convention but the balance benefits to risks prevails.

1) The INUYAMA Declaration 1990 (CIOMS Council of International Or-

[1] Some restrictions can be placed by the law on the exercise of the rights of information in the interest of the person.

ganizations of Medical Sciences) says in Art 4 that genetic tests must be performed only for the welfare of the tested person, with respect to the confidentiality of the results and accompanied by counseling to the subjects explaining the difference between carrying the gene and the genetic disease.

2) The Universal Human genome Declaration of the UNESCO[1] (9 December 1998) states that rights and dignity must be respected whatever the genetic characteristics (art 2) are. The genetic diagnosis should be performed after the evaluation of benefits and risks, with a free and informed consent, in the best interest of the subject (art 5).

3) The international declaration on Human genetics data (2003)[2]

The article 4 recognizes a special status to genetic data: can be predictive for individuals, may have significant impact on the family, offspring extending over generations, may have cultural significance, may contain information the significance of which is not necessarily known at the time of the collection...

The declaration is applicable to "medical and other scientific research" including epidemiological, especially population – based genetic studies, as well as anthropological or archaeological studies. (art 5)

It recommends the use of transparent and ethically acceptable procedures and ethics committee will be consulted with regards to establishment of standards. (art 6)

The exchanges should be organized according to the standards of the states concerned.

The art 7 recalls the principle of non – discrimination and non – stigmatization in the interpretation of findings of research on population – based genetic studies and

---

[1] Adopted on November 11, 1997 and was endorsed by the United Nation General Assembly on 9 December, 1998. The guidelines for the implementation endorsed on 16 November, 1999 by the 30C/Resolution 23.

[2] 16 October, 2003, http://portal.unesco.org.

on behavioral genetic studies.

Regarding the consent the declaration confirms in art 8 the need of a prior free informed and express consent. Some limitations are possible for compelling reason defined by domestic law, in the incompetent best interest.

As for any research, the withdrawal is possible (art 9).

The declaration also recognizes the right of the subject not to be informed of the result (art 10) and emphases the genetic counseling in an appropriate manner (art 11) and the access to own genetic data (art 13).

The respect of privacy and confidentiality is organized by the art 14, with a specific mention to the disclosure to third parties with a consent in accordance with the domestic law and the international law on human rights.

Some provisions are considered for change of purpose (art 16), the storage of biological samples (art 17), the cross boarder flow of human genetics data (art 18) and the share of benefits (art 19).

II – The situation in FRANCE

The French legislation on the analysis of the genetic characteristics is very strict. The civil code in the art 16 – 10 allows it only for medical or scientific purposes. Since 1994, the Law[1] restricts genetic analysis to scientific and medical purposes, and for evidence in courts. Article 16 – 10 of the Code Civil[2] states that: "A genetic study of the particulars of a person may be undertaken only for medical purposes or in the interest of scientific research. The express consent of the person must be obtained in writing before the carrying out of the examination, after he has been duly informed of its nature and purpose. The consent shall specify the purpose of the examination. It may be revoked without form at any time."

[1] 94 – 653 Law protects the use of the human body and its elements.

[2] 94 – 653 Law protects the use of the human body and its elements.

1) Genetic analysis for justice purposes[1]

The use of DNA for evidence in criminal investigations or in family research has replaced blood comparisons in courts. The genetic fingerprinting evidence must contain non - coding DNA segments, except the segment corresponding to the sex marker (Art 706 -54 § 5 of the Code of Criminal Procedure).

The use of DNA in criminal evidence: Forensic DNA analysis is a comparison with the DNA profile obtained from a stain of a crime scene with those from individuals in order to identify the person who possibly produced the stain[2]. For unidentified corpses, besides the anthropological methods and digital fingerprints, the comparison with the DNA of family members is the best mean to certify the identity.

In both cases, the DNA might be useful only with a reference sample to be compared with.

The civil use for paternity: civil law courts use DNA analysis when paternity is disputed or needs to be established. The consent of the subject is necessary. Article 16 - 11 of the Civil Code: "The identification of a person owing to his genetic prints may only be searched for within the framework of inquiries or investigations pending judicial proceedings or for medical purposes or in the interest of scientific research. In civil matters, that identification may be sought only in implementation of proof proceedings directed by the court seized of an action aiming either at establishing or at contesting a parental bond, or for getting or discontinuing subsidies. The consent of the person must be obtained previously and expressly. Saving an express consent given by the person during his lifetime, no identification owing to genetic prints may be effected after his death".

---

[1] The article 12 of the international declaration on human genetic data says that the data collected for the purpose of forensic medicine or in civil, criminal and other legal proceedings should be made in accordance with domestic law consistent with the international law on human rights.

[2] AM. Duguet, E. Rial, M. C Lacore, A. Cambon - Thomsen, "Forensic DNA analysis and biobanking in France in 'New chalenges for biobanks: ethics, law and governance'", Ed Kris Dierickx and Pascal Borry Intersentia, 2009, pp. 197 -208.

France has a specific regulation: the biological analysis to establish paternity is only possible on request of the judge, only for the interest of the child, and must be performed in an authorized laboratory (art 16 – 2 Civil Code and Decree 97 – 109, 6 February, 1997).

Punitive sanctions are defined in the Penal Code: one year of prison or 15000 Euros fine for genetic testing performed outside the provisions of the article 16 – 1 of the Civil Code, and without compliance to the proceedings for civil identification (article 226 – 28 Penal Code). The same sanctions are applicable for dissemination of information related to genetic fingerprints outside the judicial proceedings.

This legal procedure is very restrictive in France, but offers abound on the Internet and everybody can send samples to a foreign country and obtain a paternity test. Private societies advertise offers from 199 to 280 Euros with the results within 72 h or 5 days. Nevertheless, the result of a private paternity test performed without a judge's order cannot be used in French courts. Moreover, no genetic analysis can be performed in France without the agreement of the professional (art L1131 – 3 CSP) and the authorization of the laboratory art (1131 – 2 – 1 CSP).

2) Genetic analysis for medical and scientific purpose

a – Medical purpose

The public health code gives this definition[1] of the analysis of genetic characteristics for medical purposes: "to confirm or refute the diagnosis of genetic disease (symptoms) or search for an asymptomatic person on the characteristics of one or more genes that could eventually lead to development of a disease on the subject or his descendants".

The conditions of consent are free and informed consent given in writing after a personal medical interview. When the subject presents symptoms of the genetic disease the prescription of the analysis can be done by any physician (R 1131 – 4

[1] art R 1131 – 1 of the Public Health Code.

CSP).

For an asymptomatic person with a family's history the prescription is done by a physician working in a multidisciplinary team of genetic diagnosis, with a specific management. On the minor the prescription is allowed only if the genetic analysis leads to immediate preventive or curative measures for the minor or for his or her family members (R 1131 -5 CSP). The patient signs a certificate attesting that the information has been given and consent obtained.

The subject has the right to know the result of the genetic defect. The information tells the transmission and the personal consequences and the consequences on the family members, as well as the social and occupational consequences. The patient has the right not to know the outcomes and the refusal is recorded in the file of the person.

The result of the genetic analysis is signed by the person responsible for the analysis, and commented in the context of a clear and appropriate consultation. The subject has no obligation to inform his family that he or she is carrying a genetic disorder. This refusal cannot be the basis for a liability action in Courts (R 1131 - 14 CSP).

b - Scientific purpose

There is a difference between care and research for individual findings results. For caring the result of the genetic analysis will be an orientation for prevention or treatment. The communication of the research findings is organized by the legislation of protection of persons involved in biomedical research. Only the general results of the study are accessible to the subject (art 1122 - 1 CSP). The issue linked to incidental genetic finding susceptible to be useful for the subject is treated like all the other findings: only information that leads to medical or preventive measures should be given to the subject through the relation of the physician in charge of the biomedical research (investigator). The investigator will inform the general practitioner of the subject that genetic findings of research can be of interest and to invite the patient, if necessary, to visit a genetic counseling team in order to validate the research findings. No genetic information can be given directly to the individual without a second analysis prescribed and done in the legal frame: medical prescription and genetic counseling.

3) Specific provisions of the French law

a – Discrimination

Discrimination is a criminal offense punished by the art 225 – 1 of the Penal Code including discrimination for genetic characteristics.

Since 1995, the National Ethics Committee said in opinion on Genetics and Medicine[1] that the results cannot be used to other purposes than medical or scientific purposes. No test could be performed for insurance contracts or for access to a employment even with the agreement of the subject. The 4 march, 2002 Law on patients' rights forbids definitely the use of genetic tests for insurance (art L 1141 – 1 CSP).

b – Confidentiality of the genetic anomaly

The 78 – 17 Law Informatique et Libertés (information technology and liberties) says that informatics should not breach human rights, privacy, and individual or public liberties. The law recognizes the right for a person to be informed whether he or she is recorded in the database, to have access to the personal information stored and to oppose storage of personal data. The law makes provision for a national independent authority for data protection, (CNIL) which authorizes and controls the recording of personal data.

The subject is free to tell his genetic disease to others, but he or she must be aware of the consequences of the dissemination of this information to the family, the spouse, the employer etc.

The professional secret is general and absolute in France and the results of the

[1] Opinion of the CCNE 30/10/95 www. ccne. fr.

analysis of the genetic characteristics is given only to the patient. The physician is not authorized to disclose the professional secret even after the death of the patient, because no disclosure of the secrecy is allowed and access to medical information is impossible except when the person has, during his or her life, expressly allowed the physician to provide genetic information to the relatives, in specific circumstances.

c – Genetic counseling

Since 2010, the public health code defines the mission of the genetic counselor in the art L 1132 – 1 CSP[1]: "The genetic counselor is involved in a multidisciplinary team and practices on medical prescription and under the supervision of a qualified physician in genetics. The counselor is involved in the delivery of information and advice to individuals or families who have performed genetic analysis or who are willing to perform an examination of genetic characteristics. They can provide a medico – social and psychological follow – up. They practice in public and private health establishments authorized to conduct genetic analysis or prenatal diagnosis."

The practice of this profession is organized by the articles 1132 – 1 to 1132 – 7. There is a list of diploma allowing the practice.

d – Information of relatives[2]

Prior to conducting an examination of genetic characteristics of a person, the prescribing physician informs of the risks his or her silence exposes the family members potentially affected if a serious genetic disorder whose effects are likely to measures prevention, including genetic counseling or treatment was diagnosed.

The physician provides to this person, in a written document, the modalities of information for family members potentially affected in order to prepare for the possible transmission. When the person has expressed in writing its desire to be kept in ignorance of the diagnosis, he or she may authorize the prescribing physician to in-

---

[1] Law 2010 – 177 of 23 February, 2010.

[2] Art L1131 – 1 – 2 Public Health Code set up by the Law n°2011 – 814 du 7 Juillet 2011 – art. 2.

form the concerned relatives.

When a serious genetic defect is diagnosed, unless the person has expressed in writing its desire to be kept in ignorance of the diagnosis, the medical information provided is summarized in a document written in a fair, clear and appropriate manner, signed and delivered by the physician. The person concerned signs a document attesting that he or she has received the document.

In announcing this diagnosis, the physician shall inform the individual that one or more patient (s) associations could provide additional information on the genetic abnormality diagnosed.

The person is required to inform family members potentially concerned or, where appropriates, the legal representative since preventive measures or treatment can be offered to them.

However, the person who does not wish to inform by himself or herself the fami ly members potentially affected, may request by writing the prescribing physician to make such information. The person communicates for this purpose contact information of the interested parties. The physician informs them of the existence of a family – oriented medical information which may affects them and invites them to attend a genetics consultation. The information given to relatives does not disclose the name of the person who is the subject of examination or genetic abnormality nor risk associated with it. The information on the genetic defect will be delivered to the physician of the genetics consulting team.

e – Extension to the donors of gametes

When serious genetic disease with possible prevention measures is diagnosed on a person who has already given gametes for assisted medical procreation, the person may allow the physician to inform the center of medical assisted reproduction for the purpose of delivering information to the children born with the gametes.

Final remarks

How to protect the subject and the family of the misuse of genetic testing and the dissemination of the genetic results?

The genetic diagnosis for care or prevention is very well organized and supervised by clinicians. The new technologies allow to evaluate the predispositions and the risks, and the public is very impressed by this predictive medicine, but the scientific validity of the results needs to be assessed.

As far as some results of the predictive diagnosis remind uncertain, they cannot be balanced with potential benefits or risks. The genetic counseling is, in that respect, very relevant.

The French position is very protective for the individuals and their families and the legal frame aims to face every situation for the medical purposes: express written informed consent in any cases; genetic counseling by authorized practitioners working in a genetic clinical teams; genetic analysis performed only in laboratories with agreements, etc...

As free access to genetic tests is a real demand from individuals the French regulation should be updated to comply with the rights for free access to medical services offered in all the European Union. The control of the information provided on the direct to consumer tests on the internet should take into account the French legislation and inform the public of the consequences of the misuse of the genetic analysis.

The situation of children should be considered, having in mind the defense of the best interest which needs to be evaluated with a genetic counselor in any cases.

STUDY ON THEORETICAL BASES OF INDIVIDUAL FREEDOM FROM THE PERSPECTIVE OF PROTECTION OF GENETIC INFORMATION

CHEN Zihan[1] QI Yanping[2]

Whether focusing on the logic of life of individuals, or the individuals that are the basis of community, it is necessary to discuss the principle of individual freedom. As one of the most important political philosophies and ideologies in contemporary western society, liberalism is in interaction and mutual penetration with scientific development. It is of great significance to explore the system of rights, the philosophy of powers and the states of rights which are related to scientific development, from the perspective of the principle of individual freedom. In terms of genetic information rights, the principle of individual freedom is the primary principle of its protection.

1. The Theoretical Basis of Liberalism

Liberalism emphasizes the respect for humans, libertarian, advocates democracy, truth, criticism, and compatibility and opposes might. The starting point of liberalism is individuals. However, liberalism does not ignore the sociality of individuals.[3] Nor does liberalism exclude the concept of community, society and nation. Liberals are against all - powerful government. They regard the life, freedom and property of individuals as the basic needs to be protected. Individual freedom has always been the basic principle that is followed by liberals. In fact, as political and

[1] CHEN Zihan, Lecturer in Law School of Beijing Institute of Technology, Doctor in Law Shandong University, PhD in Biological and Medical Law University Daull Sabatier.

[2] QI Yanping, Professor of Law and Director of the Centre for Human Rights Study of Shandong University.

[3] John Horton, "Liberalism, Multiculturalism and Toleration", *Liberalism, Multiculturalism and Toleration*, Palgrave, 2001. p. 1.

aesthetic philosophy, the theoretic basis of liberalism is not single or ignescent.

In general, rationalism, individualism, social contract theory and moral pluralism are regarded as the primary theoretical basis of contemporary liberalism.[1] The basis of these theoretical values opposes each other and harmonizes with each other in the process of the protection of genetic information.

First of all, liberalism regards rationalism as its theoretical basis. The individuals' requirement of the protection of their own genetic information related to interests is also inseparable from rationalism. Certainly, the rationalism which is emphasized by the protection of genetic information rights is not the narrow rationalism, but the generalized rationalism. Rationalism itself exists in the fierce conflict of scientific research. The rationalism which is accepted by liberalism is against irrationalism, blind obedience and starting from irrationalism. At the same time, they combine experiential observation, deductive as well as inductive logic together. The rationalism which is advocated by liberalism has three characteristics. First of all, liberals believe that legitimacy lies in rational judgment, rather than relying on an individual's intuitive sense. Nor does the legitimacy lies in social customs or the order, power or authority of the persons in power. Although the development of genetic information technology is a further interpretation of the objective world, ultimately, it is different from sensory awareness. In their long – term social life, our ancestors, based on experience, summed up some quite useful genetic knowledge and strengthened them through the form of moral order. However, there is a huge difference between this and the genetic biology as discovered by genetic scientific development. The genetic information analysis as brought by genetic testing and genetic diagnosis, on the one hand, provides rational basis for the longstanding genetic experience. On the other hand, the development of genetic information technology could help the examination of the customs and ideas of social life. Secondly, as the theoretical basis of liberal-

[1] Gu Su, *Basic Concept of "Liberalism"*, Jilin Press, 2013, pp. 14 – 35.

ism, rationalism also emphasizes that the sources of social knowledge are social observation and experience. Any rational reasoning and deduction are inseparable from the observation of social life and the abstraction of sensory awareness. The development of genetic information science theory itself is a process from the observation of phenomenon to the exploration of nature. It is precisely the social observation and experience decides the objectivity and rationality of the protection of rights. Rational analysis then enhances the scientificalness and humanistic concern of the protection of rights. Thirdly, in the view of rationalists, individuals choose the rules they follow for the protection of their own rights. It is in the process of the observation of rules that the order and the rule system that the society should establish are ascertained. The protection of individual genetic information is built upon the objective needs of individuals to control their own information and protect their privacy. Meanwhile, the rules for the information to flow from individual to society are established, as an effective way to achieve information management and interests' protection. Thirdly, the starting point of liberalism is undoubtedly individualism. However, it is not a derogatory term in ideology. It does not advocate mercenary or complete egoism. Indeed, individualism establishes three basic principles, which are inseparable from the vitality of liberalism and the entire humanities that has been making contributions since modern times. Liberalism emphasizes: the establishment of human-centered experience values; human itself is the goal and human cannot be used as means to achieve other social purposes; people are equal, which means that others cannot be used as tools to achieve our own purposes.[1] In political sense, individualism does not mean anarchy. It focuses on protecting individual rights. Government is means to achieve individual interests. In a long period of time, individualism requires that market liberalism and political liberalism are in parallel, or these two liberalisms help each other to forward. However, along with the crisis encountered by liberal capital-

[1] Gu Su, *Basic Concept of "Liberalism"*, Jilin Press, 2013, p. 19.

ism, and the failure of spontaneous market regulation, individualism has been questioned once: excessive economic freedom may bring human rights abuses. However, the recession in late 19th Century, and the rampant militarism and fascism during the Second World War are regarded as an overcorrection of liberal economy. Therefore, renewed attention was placed on the respect for individual dignity and individual values which are advocated by individualism. At the same time, individualism considers that the protection of personal property is the important condition for the realization of individual value. It is like what Locke has emphasized: individual freedom is inextricably linked with life and property. In this sense, the fundamental starting point of the study of genetic information is to overcome disease. This is for the purpose of medical research, while the economic benefits of genetic information in modern society should also be respected. However, whether individual genetic information should be individualization faces the conflict with scientific research freedom and individual rights of others, within the existing institutional framework. What cannot be denied is that individualism is based on individual rights, therefore, the protection of individual genetic information is not only for the satisfaction of utilitarian purposes in political or moral sense, but for the safeguarding of the basic freedom and interests of individuals. Faced with the common concern of genetic information in the age of big data, the theoretical basis of individualism is of particular importance. This is also reflected in another aspect, which is that the right - based perspective of protection is an effective way for individual to be against interests such as community interests, society interests or national interests. Owing to the scarcity of genetic resources and the huge growing economic value of genetic information, scientific development and social welfare promotion require the obtaining of individual genetic information. However, the blind sacrifice of individual interests brought by the legitimacy of overall planning and public welfare should be avoided.

Thirdly, Social contract theory is the one of the theoretical basis of liberalism. It is straightforward to say that these theories are policy oriented, but they are very close to the medical ethics, especially in relation to resource allocation and justice issues. In 1972, Rawls' theory of justice needs to be assumed that the act of ignorance to prevent the individual from knowing their role in society whether he's a patient or a politician. Based on this kind of viewpoint, Rawls emphasizes that people will choose such a justice system, which contains two important principles. On the one hand, each person should have the freedom at the same maximum degree compatible with the other people in the group, and on the other hand, the uneven distribution of products and resources is not fair, except that the unfair distribution is to solve the most unfavorable problems.[1] Especially with the development of Rawls's social contract theory, it has a more far-reaching significance for the development of genetic information technology and social progress. The social contract theory considers that the moral criterion and the standard of judging by the society are the result of the mutual compromise. It means that the moral standards come from people's negotiation and compromise. Especially the Rawls' theory, the individual genetic information is incorporated into the framework of the legal regulation, which makes a outstanding contribution to the appearance of the personal rights. Rawls discusses a kind of principles of redistribution, which notices that the benefits not due to the personal behaviors are also important sources of injustice. Therefore, the social contract theory can make the freedom out of the historical shackles, and follow the results of people's agreement, and enjoy flexibility in the face of new things. Social contract theory provides a favorable way to solve the limitations of free capitalism market. It is more helpful for the individual to make rational choice, which extends to the social and economic fields, and it is more conducive to the protection of individual rights, social redistribution and system design.

Finally, an important theoretical basis of liberalism lies in the recognition of

[1] John Rawls, *Theory of Justice*, translated by He Huaihong, He Baogang, Liao Shenbai, Chinese Social Sciences Press, 1988, pp. 56-61.

moral pluralism. Liberalism does not set an only consensus or a supreme moral. From the starting point of the individual's respecting, it is necessary to recognize the differences of individual value judgment. Neutrality is an important character in the face of different moral values for liberalism. Therefore, liberalism thinks that the important rules of the free society, such as the idea of justice can be deduced from different moral values. So liberalism holds that justice is independent of good. The liberalism protects the individual freedom at the maximum level to act as the only judge of pursuing their own happiness. Freedom is not involved in realizing freedom belong to others, it should be respected. At the same time, pluralism of moral of liberalism doesn't mean nihilism, but represent a inclusive ability of liberalism. Liberal moral pluralism and value of universality are not in conflict with each other and liberalism treating social rules independent of the good and out of the moral conception, let universal value be formed. In this sense, liberalism is the foundation of rights protection. The development of the society, especially the regionalization and globalization has accelerated the integrity of whole society. From the perspective of human society development, we have a huge society. The social integration will inevitably enjoy moral pluralism. From this point of view, we have not built up a whole society around a single moral concept or a unified ideology. People follow different moral rules. According to Professor H. Tristram Engellaardt's point of view, based on the different moral situations, people are split into moral friends or moral strangers, two kinds of group relations.[1] The moral friend, in the face of the dispute can solve them through a common understanding of the moral standards or by appealing to the same moral authority. Instead, moral strangers based on having no common moral standards or doing not agree with the same moral authority, can not be like a friend to tackle the problem, but they can resolve the dispute by mutual agreement. This

[1] See H. T. Engelhardt, *The Foundation of Biological Ethics*, translated by Fan Ruiping, Peking University Press, 2006, p. 13.

situation, in the doctor – patient relationship, has a very clear manifestation. Based on asymmetry of medical information and expertise experience, physicians and patients can not be moral friends, but only as moral strangers and build the same moral environment to salve the problems.

Doctors' morality is different from the patient's morality, such as the doctor is pursuing the health and happiness for patients, but they can not deny the right to give birth to a baby with defects and disease already diagnosed of parents. But these morals must follow the general good in today's society. When doctors no longer pursue the benefits of patients, but simply pursue their economic value, it will no longer be a moral conflict, but the moral decay. It does not involve the moral pluralism negotiation, but should bear the responsibility brought by the lack of morality. In addition, the moral conflicts between doctors and patients can be resolved through consultation. Doctors do indeed have a bias in their own interests, but the informing specific situation is the premise into an agreement, when a patient has a risk or conservative personal preference. Each genetic information technology has different affection on the future of life and health of body for patients in the diagnosis and treatment, and the scientific research are trying to conquer the risk brought by the application of genetic information. So the establishment of discussions on the explanations and the individual cognition become an important principle of the protection of genetic information rights. Therefore, moral friends are with the common substantive moral, while moral strangers are with common procedural moral. This kind of procedural morality is derived from the discussion. The principle of informed and consent followed by the users and providers of genetic information is the establishment of such a kind of procedural morality.

2. The Value of Individual Freedom

In political philosophy, there is no more important and more complex than the concept of "freedom" which is discussed by countless scholars. In Montesquieu view, freedom is the right to do everything allowed by law.[1] This famous thesis contains three important aspects: Firstly, freedom means the right of choice. If you have no choice, you do not have the freedom. Secondly, freedom is not free, and the freedom by liberalism is the restriction of the freedom in the system and thirdly, the freedom talked by the liberalism is to establish what limits are legitimate and reasonable. Personal genetic information is closely related to personal life and health, family relationship and other interests. Therefore, the individual has the right to choose to know the status of genetic information, and the right is established by international law.[2] The freedom to obtain personal information, not only means that you are willing to know the status of genetic information but also means that not to know the status of individual genetic information. Even in the doctor – patient relationship, this choice which is not to know effectively prevents the professionals' paternalistic decision, and reflects the autonomy of the parties. The autonomy of genetic information is also reflected in the right of processing of genetic information. "Man is born free and everywhere he is in chains."[3] The exercise of genetic information rights is also influenced by many factors. In particular, it is difficult to understand the status of genetic information without the professional interpretation. Therefore, the protection of genetic information is largely responsible for the conflict between the

---

[1] C. L. Montesquieu, *The Spirit of the Laws* (French: De l'esprit des Lois, originally spelled De l'esprit des loix; also sometimes called The Spirit of Laws, in which Montesquieu distinguishes this view of liberty from two other, misleading views of political liberty. The first is the view that liberty consists in collective self – government—i. e. that liberty and democracy are the same. The second is the view that liberty consists in being able to do whatever one wants without constraint. Not only are these latter two not genuine political liberty, he thinks, they can both be hostile to it).

[2] UNESCO, International Declaration on Human Genetic Data, October 16th, 2003, Article 13.

No one should be denied access to his or her own genetic data or proteomic data unless such data are irretrievably unlinked to that person as the identifiable source or unless domestic law limits such access in the interest of public health, public order or national security.

[3] L' homme est né libre et partout il est dans les fers. Du contrat social ou Principes du droit politique, which English name is Social Contranct or Principles of Political Right by Jean Jacques Rousseau, translated by G. D. H. Cole, public domain. 1762, p. 4.

ownership rights and derivative rights, and for clarifying the boundaries of the rights. Therefore, the freedom has two functions, Firstly, to achieve their own control, that is, independent and secondly, to coordinate conflicts, that is, reasonable.

In the liberal view, the individual has freedom of choice. Thereforc, they do not think that freedom should be given to a small number of people to be implemented, such as the existence of the feudal patriarchal system or the privileged class. The choice does not mean that all is in line with the moral and the good choice, the so - called freedom, should be a neutral action. Freedom contains the meaning that people are allowed to make mistakes, correct and improve. In most cases, people with professional skills, such as scientific research institutions, medicines company, researchers, medical workers, have more material and professional knowledge than those who have individual genetic information. In some special circumstances, paternalistic decision is also likely to be the best choice for the individual or to improve the well - being of the citizens, but this does not constitute a reason to prevent the individual choice.

At the same time, the liberalism distinguishes the differences between the conditions for realization of freedom and limitation for realization of freedom. The freedom without limitation and the ability to carry on one activity are different. Law guarantees that people can freely carry on market transactions, but can not meet the desire of everyone to buy. For example, anyone has the right to know the status of its own genes, but does not mean to obtain the individual's genetic testing report with free of charge.

Even in current times, genetic testing technology is not universal, and still need to be paid higher costs. Freedom is in need of legal protection and permitted by laws which ensure the majority to realize their freedom. In Rock's view, the law is the guarantee of freedom, not the deprivation or restriction of freedom. The law protects the most interests rather than a small number of freedom or privilege. The law is based on equality and non discrimination. At the same time, the law gives a certain limit to the individual to make the individual have the ability to achieve the full and reasonable freedom, then the limit is a means while protecting freedom is the purpose. Although the law provides for the space of individual action, it still retains the right to personal choice rather than providing a single active path.

As a master of liberal thought, Berlin's discussion has a great significance. His point of view mainly includes three aspects. Firstly, Berlin thinks there is the distinction between negative freedom and positive freedom. Secondly, people's desire are in thousands of types but they are not completely different. And thirdly, freedom and other value does not equal, the loss of freedom can not get remedy by the other kinds of values. Although for the size of the scope of freedom the classical political philosophers have different views, they basically agree with the personal freedom is not completely unrestricted. Complete freedom may lead to absolute chaos, which may make some people thoroughly oppressed to meet another endless desire. And it is contrary to the idea of freedom and equality advocated by modern times. If genetic information circulation is without limit, as the sale of organs, people with inferior position in economy, technology, and information are likely to be active or passive to become vassals of the dominant group, regardless of whether the dependency relationship will hurt a person's rights of life and health or not.

3. The Principle of Individual Freedom Is Reflected in the Autonomy

Autonomy literally means self – controlling. Autonomy is valuable because it is able to form our personalities through independence, it is also important for what we are determined and why we have dignity. Autonomy in the protection of genetic information rights is very obvious. In the process of the protection of genetic information, we need a person who has the ability to recognize and control his own decision, must be able to grasp the risks and potential benefits, and can eliminate the interference of the outside world. Specifically, in the case of scientific research and commercial interests, the conditions for genetic detection and gene diagnosis that must meet the above requirements and rule out the obstruction of third people can be called autonomous. In general, it is urgent to establish the individual freedom principle of genetic information. On the contrary, the principle of non compliance, will lead to many adverse consequences. A case in the history of genetic medicine has become an example for bitter experience. In 1999, only 18 year – old, Jesse Gelsinger, because of suffering from an ornithine transcarbamylase deficiency, went to University of Pennsylvania Medical Institutions for treatment. The doctor James Wilson enjoyed the related patents of testing adenovirus carrier transfer of gene, and as the owner of the pharmaceutical companies, he preferred Jesse to receive the gene therapy. Wilson did not explain to the participants that a test sample of a monkey had died in the previous experiment, that is, there was still a huge risk for the human trial. Unfortunately, Jesse was killed in few hours after receiving an injection of a recombinant adenovirus vector, which was rejected by immune rejection. After investigation, Jesse, although had been over 18 years of age, and in the treatment would be sober, it is obvious that the subjects can not grasp the risk of the genetic test, and got induced suggestion by the doctors and finally made the decision to accept the genetic test. The attending physician and scientific research institutions are clearly contrary to the principle of respect for autonomy. According to Jesse Kissinger's case, it can be seen that respect for autonomy is an active duty, which requires the active cooperation of other subjects.

DIRECT - TO - CONSUMER GENETIC TESTING ON THE INTERNET: COMPARISON BETWEEN THE LEGAL FRAMEWORKS IN FRANCE AND IN CHINA

WU Tao[1], Gauthier Chassang[2][3]

Abstract:

Direct - to - consumer genetic testing (DTCGT) is a quite new commercial practice consisting in the sale of genetic testing services over the internet, directly to consumers, without necessarily involving qualified medical staffs and genetic counselling.

Such a business has been criticised in numerous way due to the important legal ethical and social issues it entails. Some states have adopted more or less direct legal frameworks about this phenomenon or are in a process for regulating this economical field.

This article aims to identify, analyse and compare existing regulation of DTCGT in Europe, and particularly in France, and in Asia, particularly in China, in order to figure out the existence of adapted legal provisions and analyse the adopted regulatory approaches.

Through the analysis, we will take examples from other countries, like with the regulatory activity in the United States of America (USA), and examples of two DTCGT companies offering their services to Chinese citizens in order to show the diversity of regulatory positions and initiatives, and, where necessary, highlight perspectives regarding the current regulatory needs.

[1] *Department of Health Management*, Xi'an Medical University, Shaanxi Province, China.

[2] Inserm, UMR1027, Toulouse, F-31062, Midi-Pyrénées, France.

[3] Université Toulouse III, UMR1027, Toulouse, F-31073, Midi-Pyrénées, France.

Keywords: Genetic Testing/Direct – to – consumer/Online Genetics/Law/Internet

The completion of the Human Genome Project in 2003 and the rapid advances in human genomic analyses associated with the decreasing cost of related genetic testing technologies provided new opportunities to develop innovative health services. In a dynamic of patient – centred healthcare and of personalised/stratified medicine intended to better integrate Deoxyribonucleic Acid (DNA) testing tools and methods into the clinical and research settings for improving early prevention, diagnosis and efficient treatments of diseases, other activities giving a central role to personal autonomy developed, such as the "Direct – To – Consumer Genetic Testing" (DTCGT) activities.

DTCGT refers to the commercial offers of human biological samples' DNA analysis services directly sold to consumers via the internet, television, or other marketing means, without necessarily involving qualified health professionals. With DTCGT, genetic testing is not anymore dedicated to patients and healthcare professionals but opens now to consumers. The tests offered online concerns a wide range of health conditions but also included non – health related genetic tests like ancestral or paternity genetic testing. Within this article, we only tackle health – related DTCGT, even though the non – health related genetic tests also calls for ethical, legal and social reflections[1]. DTCGT services include genetic testing for both monogenic disorders and for genetic variants associated with common complex or multifactorial diseases (susceptibility variants) as well as testing for other traits like sub-

[1] E. g. Martine Betti – Cusso. Secrets de famille: ces tests AND qui dévoilent la vérité. LeFigaro, Société, published online the 3rd of January 2014, Available at http: //www. lefigaro. fr/actualite – france/2014/01/03/01016 – 20140103ARTFIG00249 – secrets – de – famille – ces – tests – adn – qui – devoilent – la – verite. php (Accessed on 1st of September, 2014).

stance reactivity or dependence[1][2]. Most of the tests have a predictive value. While legal definitions of genetic testing and genetic tests vary according to jurisdictions[3][4], DTCGT companies offers cover the whole spectrum of genetic tests. For the purpose of this article, we will use the definitions of genetic testing given by the European Commission Independent Expert Group of 2004. The group provided a general definition of genetic testing as "any test that yields genetic data[5]", whatever the nature of the test, the methods or the technology used, provided that it is revealing information about the genetic characteristics, germ – line or somatic information[6]. The Expert Group further describes genetic testing by differentiating diagnostic testing from predictive testing. The latter category is at the heart of the DTCGT offers that provides consumers with an attractive mean to identify their personal risks to develop a disease, to have an adverse reaction to a medicinal product (pharmacogenomics) or to have a specific metabolic reaction regarding food products or com-

---

[1] Christina R. Lachance, Lori A. H. Erby, Beth M Ford, Vincent C. Allen Jr and Kimberly A Kaphingst, "Informational content, literacy demands, and usability of websites offering health – related genetic tests directly to consumers", *Genetics in Medicine*, 2010, pp. 304 – 312. See Table 1 – Categories of health conditions for which sites offered testing.

[2] Genetics & Public Policy Center, List of companies and tests offered, August 2011, Updated January 2012. Available at the following address (Accessed on 1st of September, 2014): http: //www. dnapolicy. org/images/reportpdfs/NewMethodsForDTCTable_ updated_ Jan2012. pdf

[3] Varga O., Soini S., Kääriäinen H., Cassiman J. J., Nippert I., Rogowski W., Nys H., Kristoffersson U., Schmidtke J., Sequeiros J., "Definitions of genetic testing in European legal documents" *J Community Genet*, 2012 Apr; 3 (2): 125 – 41. doi: 10. 1007/s12687 – 012 – 0077 – 1. Epub the 26 January, 2012.

[4] EuroGentest, Orsolya Varga and Jorge Sequeiros and al. Definitions of Genetic Testing in European and other Legal Documents, 10th (final) draft version – 16 June 2009. Available at http: //www. eurogentest. org/fileadmin/templates/eugt/pdf/BackgroundDocDefinitionsLegislationV10 – FinalDraft. pdf (Accessed on 1st of September, 2014).

[5] European Commission, the Independent Expert group. Ethical, legal and social aspects of genetic testing: research, development and clinical applications. ISBN 92 – 894 – 7324 – X, p. 24. Brussels, 2004.

[6] Using one or more of these methods, genetic testing can yield different types of information such as:

- Confirmation or exclusion of the diagnosis of a specific disease;
- The magnitude of the risk of developing a disease or of adverse reactions to treatments and environmental factors;
- The magnitude of the risks that biological descendants will inherit a defect.

ponents (nutrigenomics). Genetic testing services are provided online to asymptomatic and healthy consumers but also to patients, at a presymptomatic stage and even at a prenatal stage. The offer also potentially concerns children. Basically, the consumer orders the test for one or more conditions he selected and pays a price online. Then, he receives a device for gathering human biological samples, usually saliva, to send back to the firm that will process the samples and analyse it. Once the results generated, the consumer is either invited to access them through the company's website or he only and directly receive an email containing the raw or lowly interpreted genetic testing results. Each step of DTCGT process sparkled ethical, legal and social concerns and numerous authors scrutinising websites warned about the potential harms and negative consequences of this commercial practice. Most of observers call for appropriate oversight and regulation of this e - market that could weigh 233. 7 US dollars[1] by 2018. While most of the DTCGT companies are established in the USA, they started to establish in other countries such as in Europe (e. g. EasyDNA [UK]; HairDX [Italy]; GenTest [Czech Republic]; GENEPLANET [Slovenia]) and in Asia, as in China (e. g. BGI DX). Even if they are established elsewhere, DTCGT companies can always provide services to Europeans or Chinese e - consumers. Therefore, in the absence of specific international regulation of DTCGT, it appears interesting to study the state - of - art of national relevant regulations that frame or would be of interest for framing this activity. What is the status of DTCGT regulation in France, Europe, and China? We will analyse and compare the relevant regulations and related ethical issues in France and Europe (I), as well as in China (II). We will punctually refer to other models, like the USA one.

[1] Future of Direct - to - Consumer (DTC) Genetic Testing Market Remains Fraught with Challenges, According to New Report by Global Industry Analysts, Inc. e - pub 8 August, 2012. See the following link (Accessed on 4 September, 2014). http: //www. prweb. com/releases/DTC_ genetic_ testing/direct_ to_ consumer_ tests/prweb9780295. htm.

1. Direct – to – consumer genetic testing and the law in France and in Europe

1. 1. An activity that is legally prohibited in France

Most of national laws in European countries regulate genetic testing practice for health purposes and provide relevant rules regarding DTCGT. Three models [1] can be distinguished: some countries allow the provision of DTCGT (E. g. UK, Belgium), some others restrict it (E. g. Netherlands) and others forbid DTCGT (Germany, Portugal, Switzerland and France).

In France, pre – and post – natal genetic testing practices are regulated since the 1990's, notably through the Bioethics Laws of 1994 [2], 2004 [3] and 2011 [4]. All the legislative measures constitute a specific regime regulated by the articles of the Civil Code, the Public Health Code and the Penal Code. According to Article 16 – 10 of the French Civil Code, "the examination of the genetic characteristics of a person can only be performed for medical or scientific research purposes, or based on a judicial decision." This provision excludes possibilities to access genetic testing for nonmedical or non scientific research necessities, for personal curiosity or "for fun". In France, genetic testing is a particular informational tool to be used by

---

[1] Pascal Borry, Rachel E. van Hellemondt, Dominique Sprumont, Camilla Fittipaldi Duarte Jales, Emmanuelle Rial – Sebbag, Tade Matthias Spranger et al. Legislation on direct – to – consumer genetic testing in seven European countries. *European Journal of Human Genetics* (2012), 1 – 7, online publication, 25 January, 2012; doi: 10. 1038/ejhg. 2011. 278.

[2] Loi n° 94 – 548 du 1er juillet 1994 relative au traitement de données nominatives ayant pour fin la recherche dans le domaine de la santé et modifiant la loi no 78 – 17 du 6 janvier 1978 relative à lìnformatique, aux fichiers et aux libertés, NOR: RESX9200045L, JORF n°152, 2 Juillet, 1994, p. 9559.

Loi no 94 – 653 du 29 juillet 1994 relative au respect du corps humain, NOR: JUSX9400024L, JORF n° 175 du 30 Juillet, 1994, p. 11056.

Loi n° 94 – 654 du 29 Juillet, 1994 relative au don et à l'utilisation des éléments et produits du corps humain, à l'assistance médicale à la procréation et au diagnostic prenatal, NOR: SPSX9400032L, JORF n°175 du 30 Juillet, 1994 p. 11060.

[3] Loi n° 2004 – 800 du 6 août 2004 relative à la bioéthique, NOR: SANX0100053L, JORF n°182 du 7 Juillet 2004, p. 14040, texte n° 1.

[4] Loi n° 2011 – 814 du 7 juillet 2011 relative à la bioéthique, NOR: ETSX1117652L, JORF n°0157 du 8 Juillet, 2011 p. 11826, texte n° 1.

health professionals or scientists only in certain circumstances, as a support for medical decision – making. Clinical genetic testing supports and facilitates the diagnostic, prevention and treatment of diseases with established or suspected genetic features. In research, genetic technologies serve to discover new knowledge about the diseases and to explore ways to enhance existing clinical practices. Article 16 – 10 of the French Civil Code also requires getting the written, specific, explicit and informed consent from the person to be tested before performing the test. This restrictive approach to genetic testing in general is completed by the Article 226 – 28 – 1 of the French Penal Code, created by the last Bioethics Law of 2011 [1], stating that "The fact, for a person, to solicit an examination of its genetic characteristics or of those of a third [...] outside the conditions planned by the law is punished by a 3750 euros fee". Thus, the legislator decided to prohibit DTCGT by forbidding access to genetic testing outside the health system. This very strict regulatory model sparkled some critics and deserves more attention regarding its rationale and its real impact on DTCGT.

The rationale of the prohibition seems related to the will to avoid that citizens access to paternity genetic testing which is only available under judicial decision in the French legal order. However, the risks and benefits of health – related DTCGT has been scrutinised in France through the literature [2][3][4] and through the works of several instances such as the French National Consultative Ethics Committee

[1] Loi n° 2011 – 814 op. cit. Article 6.

[2] Gauthier Chassang, Emmanuelle Rial – Sebbag. Anne Cambon – Thomsen. Les aspects éthiques, légaux et sociaux des tests génétiques en accès libre. Séminaire d'actualité de droit médical. Droits des patients, mobilité et accès aux soins, Ve forum des jeunes chercheurs. Juin 2011, pp. 197 – 209.

[3] Anne Cambon – Thomsen. L'information génétique dans la société de l'information. Revue Politique. Bioéthique, e – pub 28 Février, 2012.

[4] See the compilation of articles in Revue Générale de Droit Médical, Acte Table Ronde. Accès aux tests Génétiques en Europe: Droit et protection des utilisateurs. RGDM n°42, pp. 12 – 54, March 2012.

(CCNE)[1][2], the French National Institute of Health and Medical Research[3] (INSERM), the French Biomedicine Agency[4], the French Parliamentary Assessment Office for Scientific and Technological Choices[5] (OPECST) or in the frame of research projects[6]. A constant reflection about this direct approach of consumers using marketing methods has been performed in parallel to existing reflections around the existing and future clinical and research uses of genetic testing, regarding the development of the so-called "precision medicine", "personalised/stratified medicine". With DTCGT, the reflections organised around the new model conveyed by the companies based on personal autonomy and on predictive medicine, as well as on the impact of DTCGT market regarding existing practice of genetic testing enshrined within the traditional health system. At the European level, the Science and Technology Options Assessment board (STOA) of the European Parliament published a study report[7] analysing 32 DTCGT websites' offers and the European Society of Human Genetics (ESHG) adopted specific recommendations[8]. Several

[1] CCNE, Avis n°86, Problèmes posés par la commercialisation d'autotests permettant le dépistage de l'infection VIH et le diagnostic de maladies génétiques. 4 November, 2004.

[2] CCNE, Avis n°46, Génétique et medicine: de la prediction à la prevention, 30 October, 1995.

[3] INSERM, Expertise Collective. Tests Génétiques - Questions scientifiques, médicales et sociétales. Collective Expertise; Genetic tests - Scientific, medical and societal issues. ISBN 978-2-85598-870-5, November, 2008. See notably p. 36 and following for DTCGT.

[4] E. g. Professional Seminar organized by the Council of Europe, the French Ministry of Health and the Biomedicine Agency. Tests génétiques en accès libre et pharmacogénétique - direct-to-consumer genetic testing and pharmacogenetics, Paris, 2 October, 2007.

[5] French Parliament, OPECST Report n° 1724. Les progrès de la génétique, vers une médecine de précision? Les enjeux scientifiques, technologiques, sociaux et éthiques de la médecine personnalisée (MM. Alain Claeys et Jean-Sébastien Vialatte), 22 January, 2014. Available at the following address: http://www.assemblee-nationale.fr/14/pdf/rap-off/i1724.pdf (Accessed on 3 September, 2014)

[6] TeGALSI Project - Tests Génétiques en Accès Libre Sur Internet - Direct-To-Consumer Online Genetic Testing, funded by the French Institute for Public Health Research (IReSP), 2010-2013.

[7] European Parliament, STOA, Study Report. Direct-to-consumer Genetic testing. Study IP/A/STOA/FWC/2005-28/SC32 & 39. PE 417.484. November 2008.

[8] ESHG, Statement of the ESHG on direct-to-consumer genetic testing for health-related purposes has been published in the European Journal of Human Genetics, 2010, 1-3.

studies also tried to understand the DTCGT users' motivations and feelings [1] as well as the commercial strategies [2] underlying these offers, some others studied health professionals' attitudes [3] regarding DTCGT. In summary, the following important ethical and legal issues related to the DTCGT offers have been highlighted:

- Problems related to the reliability of the commercialised tests, analytical validity, to the clinical validity [4] and clinical utility, [5] of the results;
- Problems related to the quality of the laboratories and to the necessary individualised medical supervision by qualified professionals and to an independent genetic counselling to interpret the results, risks of psychological harms for the patients or of physical harms due to inappropriate autonomous medical decision – taking;
- Problems related to the respect of the consumers' right to clear, intelligible and objective pre – test information, to the free informed consent and to the validity

---

[1] Pascal Ducournau, Claire Beaudevin. Know your genes. The marketing of direct – to – consumer genetic testing. Of deterritorialization, healthism and biosocialities: the companies' marketing and users' experiences of online genetics, *Journal of Science Communication*, JCOM 10 (3), September, 2011.

[2] Ducournau P., Gourraud P. – A., Rial – Sebbag E., Bulle A., Cambon – Thomsen A. Tests génétiques en accès libre sur Internet. Stratégies commerciales et enjeux éthiques et sociétaux. Journal Médecine/ Sciences, n°27, 2011, pp. 95 – 102.

[3] Howard HC, Borry P. Survey of European clinical geneticists on awareness, experiences and attitudes towards direct – to – consumer genetic testing. Genome Med. 2013 May 22; 5 (5): 45. doi: 10.1186/gm449. 2013.

[4] "Scientific validity", refers to the way in which the test measures the characteristic it is designed to identify. In particular, this concept includes the capacity that the test will be positive if the genetic characteristic is present (analytical sensitivity), and negative if it is absent (analytical specificity).

"Clinical validity" of a test is to be understood as corresponding to a measurement of the accuracy with which the test identifies or predicts a clinical condition. It is defined in terms of clinical specificity, sensitivity and predictive value.

"Clinical utility" is to be understood by the value of the test results in guiding the person concerned in his or her choices regarding prevention or therapeutic strategies.

Council of Europe, Additional protocol to the Convention on Human Rights and Biomedicine, concerning Genetic Testing for Health Purposes, CETS No. 203, Explanatory Report, Article 5, points 48 – 49, and Article 6, point 57, Strasbourg, 27 November, 2008.

[5] "Clinical utility": the impact of a test on clinical decision – making, and medical or health economic outcomes". See, European Commission, the Independent Expert Group, op. cit. 2004, p. 19.

of commercial contracts[1];

• Problems related to the websites' regular evolutions (e. g. changing terms and conditions);

• Problems related to the existence of an efficient right to withdraw consent;

• Problems related to the creation of human biobanks and commercialisation of the biological resources gathered from DTCGT (e. g. As it happened with the closure of the famous DTCGT companies Navigenics, deCODEme and the selling of DTC - based biobanks[2].);

• Problems related to privacy protection, identification of the source person as the consumer and illegal testing online, reuses of the ressources, unsolicited findings' management, data confidentiality and security;

• Problems related to the public authorities' oversight of activities;

• Societal problems: risks of stigmatisation, eugenism[3], social discrimination, "healthsism", "geneticisation", beliefs in "genetic determinism" and the rise of nefast "biopolitics" or "biofinance[4]".

Therefore, according to the French law, access to health - related genetic testing is neither "free" nor DTC but circumstanced to medical necessities and only available to patients through the traditional health system ensuring the quality of the

---

[1] E. g. Elsa Supiot. Le consommateur de tests génétiques, un patient avisé ou berné ? Revue des contrats, 1er October, 2009 n°4, p. 1573.

[2] A $ 415 million sale of deCODE Genetics to Amgen. Amgen will not anymore provide DTCGT.

In July 2012 the company Life Technologies acquired Navigenics. Financial and other terms of the acquisition were not disclosed.

[3] E. g. "A patent secured by California - based 23andMe gives the company the rights to match the genetic profile of would - be parents with donor sperm and eggs". See Gautam Naik. Designer Babies: 'Patented Process Could Lead to Selection of Genes for Specific Traits. Frontiers of Genetic Enhancement Continue to Advance. The Wall Street Journal, US news. 3 October, 2013.

[4] Genomics Law Report. As deCODE Departs, 23andMe Reloads. Brief posted by Dan Vorhaus on December 11, 2012.

medical biology act[1], of the test and of the individualised ongoing or clinical taking in charge of the users. In the traditional health system, access to health – related genetic testing necessitates a medical prescription[2] from a qualified physician[3], geneticist or not. Good practices in prescribing and implementing health – related genetic testing are fixed by the law[4]. In a face – to – face interview the person is informed about the necessity and potential consequences of having such a test, particularly regarding family or pregnancy of the patients, the possibilities to prevent the diseases or to treat it based on the test results, and is eventually oriented toward genetic counselling service. Testing of asymptomatic minor must be particularly justified. If necessary, interdisciplinary teams of genetic counsellors[5] provide further information and support to the patient in a non – directive way, to facilitate the understanding of the testing process, its rationale and potential results (pre – test counselling), to empower patient in decision – making (informed consent). After the test as well, genetic counsellors will explain the implications of the results and perform individual ongoing (post – test counselling). After the provision of the necessary information[6], physicians must get the specific written consent of the patient or its legal representatives. The communication of the results from the laboratory to the prescriber physician and the patients is also regulated[7] in a way that allows both further interpretation and validation of the results as well as the respect of medical secret

[1] Article R. 1131 – 2 and related exclusions from Article L. 6211 – 1 of the French Public Health Code.

[2] Article R. 1131 – 4 and R. 1131 – 5 of the French Public Health Code.

[3] Articles R. 1131 – 6 to R. 1131 – 12 of the French Public Health Code ("agrément" of prescribers).

[4] Arrêté du 27 mai 2013 définissant les règles de bonnes pratiques applicables à l'examen des caractéristiques génétiques d'une personne à des fins médicales, NOR: AFSP1313547A, JORF n°0130 du 7 Juin, 2013 page 9469, texte n° 14.

[5] On qualifications and role of genetic counsellors, see Articles L. 1132 – 1 to L. 1132 – 7 of the French Public Health Code. For the EU genetic counsellor's activity in France, see Articles R. 1132 – 4 and 5.

[6] Arrêté du 20 Juin, 2013, op. cit. Point 3.

[7] Articles L. 1131 – 1 – 3 and R. 1131 – 19 of the French Public Health Code.

and confidentiality. Since the Bioethics Law of 2011, a specific procedure [1][2][3] must be respected to communicate genetic information about a preventable/treatable disease to family members. The medical biology laboratories involved must have been authorised [4] by the Regional Agency of Health and annually report their activities to the Biomedicine Agency. Laboratories must mandatorily be accredited by the CO-FRAC [5], then, they are officially recorded [6]. The professionals performing the test must be accredited ("agrées" in French) by the Biomedicine Agency, for 5 years.

Concerning the prohibition principle, DTCGT companies and convinced consumers tends to highlight the benefits of the direct access to genetic services as a mean for decreasing inequalities to access to health services. This argument seems not to be viable for several reasons. Firstly, everybody does not have access to the internet. Secondly, DTCGT remain commercial offers, e – consumers have to pay for a service that is free for patients within the traditional health system. Thirdly, e – consumers have to face marketing strategies they are not always aware of. These strategies make genetic testing attractive by also conveying messages that can be misunderstood by the consumer or interpreted in a way that reinforce false ideas of "ge-

---

[1] Article L. 1131 – 1 – 2, Articles R. 1131 – 20 – 1 to R. 1131 – 20 – 5 of the French Public Health Code.

[2] Décret n° 2013 – 527 du 20 juin 2013 relatif aux conditions de mise en œuvre de l'information de la parentèle dans le cadre d'un examen des caractéristiques génétiques à finalité médicale, NOR: AFSP1311381D, JORF n°0143 du 22 Juin, 2013 page 10403, texte n° 4.

[3] Arrêté du 20 juin 2013 fixant le modèle de lettre adressée par le médecin aux membres de la famille potentiellement concernés en application de l'article R. 1131 – 20 – 2 du code de la santé publique, NOR: AFSP1311382A, JORF n°0143 du 22 Juin, 2013 page 10405, texte n° 7.

[4] Articles R. 1131 – 13 to R. 1131 – 18 of the French Public Health Code.

[5] Article L. 1131 – 2 – 1 of the French Public Health Code. COFRAC: Comité Français d'Accréditation.

[6] Biomedicine Agency, List of the French medical biology laboratories: http: //www. agence – biomedecine. fr/Etablissements – ou – laboratoires.

netic exceptionalism[1]" or "genetic determinism[2]". As an example, most of consumers might believe that the samples and data they paid for are destroyed after analyse and return of results, but in reality companies create biobanks including databases that they will reuse or commercialise in the future for their own profit. Additionally, the applicable law to these activities that go further than the sole provision of the initial expected service for some consumers, could be ruled by foreign laws, maybe less protective than the French one. Such details are generally unknown from the consumers which do not read or do not properly understand the thousands words composing the terms and conditions of the offers and naively trust the company as normal health service providers, just as in the health system, what they are not exactly. DTCGT actors are first of all businessmen. Most importantly, testees could misunderstand the results of the tests and be harmed by the "misinterpreted or erroneous predictive health information which overstates the role of genetics in developing common diseases. This may result in delays in seeking proper medical advice (or seeking unnecessary medical treatment) or making expensive and unproven dietary or lifestyle changes[3]". Therefore, e – consumers are put at risk to undertake payable procedures they do not need and whose they do not properly understand the scope. Thus, DTCGT tends to increase inequalities on technological, monetary and literacy grounds, which could create health risks and a two – speed health system damaging social solidarity principle. Furthermore, by forbidding access to this offer the French legislator does not affect the availability of genetic testing for health purposes but privilege access through the traditional, secured existing health system. Criticisms emerge regarding the choice of the French law to adopt a paternalistic

---

[1] "Genetic exceptionalism": The belief that the particular nature of genetic information gives rise to greater risks or particular risks that are different from other health – related risks. European Commission, the Independent Expert Group Report, 2004, op. cit.

[2] On this notion, see the INSERM Collective Expertise of November 2008 op. cit. Part 1, pp. 3 – 122.

[3] Human Genetic Commission (UK), "Common Framework of Principles" for direct – to – consumer (DTC), 4 August, 2010.

model forbidding the access to DTCGT services, thus depriving citizens form a part of their autonomy in their own health management. This position seems to contradict fundamental patients' right to access information about their health, as recalled by the ESHG[1] and fixed by the French Patients' Rights Act[2] of 2002 and Article L. 1111 –2 and 7 of the French Public Health Code. But these latter measures are limited to the access of personal health information gathered in a medical or a scientific research process undertaken by the sole health professionals and establishments set up as a part of the health system which comply with all the legal requirements for practicing genetic testing and processing health genetic data. Thus, the French legislator restricted citizens' freedom to access genetic data through services that does not meet the standards of good practices, including "for funny" purposes. By doing so, the French legislator respected the ESHG Statement mentioning that "Individuals are entitled to health information and genetic information about themselves. However, this right to know must be exercised with due respect for the need to protect the same individuals from inappropriate genetic information and testing".

Ultimately, we can note that the Article 226 –28 –1 of the Penal Code also forbids the use of DTCGT services by health professionals but does not preclude the provision of services to professionals, like prescribers[3] of tests. The French law aims to protect the public interest and to safeguard social justice, health system and patients' rights to quality health services in the absence of true necessity or consensus about the risks and benefits of DTCGT.

Concerning the impact of the prohibition, the law does not explicitly forbid the establishment of DTCGT companies on the French territory provided that they comply with other legal requirements described above and do not target/supply the French

[1] ESHG, Statement on direct – to – consumer genetic testing for health – related purposes, 2010, p. 1. op. cit.

[2] Loi n° 2002 – 303 du 4 mars 2002 relative aux droits des malades et à la qualité du système de santé, NOR: MESX0100092L, JORF du 5 Mars, 2002 page 4118, texte n° 1.

[3] On a direct – to – providers/prescribers/physicians (DTP) model.

public. A side - effect would be to overcharge citizens with personal responsibility and to prevent them to claim in front of French or European jurisdictions in case of abuses because they accessed DTCGT by infringing national law. This would result in an absence of protection. Prohibition could also urge health systems to provide any necessary genetic testing in order to create a sort of monopoly reducing interests in DTCGT. Regarding law enforcement, as other Internet regulations, it is hard to believe that it can prevent French citizens to buy DTCGT offered by foreign companies if these latters do not consider different enforceable legislation through their offers. The use of internet weakens the impact of such national measure and, in the absence of communication, education of citizens, control means, international convention or specific guidelines for operators, we can guess that the effectiveness of such prohibition will be low. However, this strict position should diminish potential negative impacts of DTCGT on health systems' economy and consumers' rights and wellbeing. At least, it makes clear that DTCGT is deemed as endangering consumers and society in many ways and is thus disapproved.

1.2. An activity internationally framed in the context of the Council of Europe and that tends to be further considered with the European Union legal reform on in vitro diagnostic medical devices

The French prohibition of DTCGT remains a quite unique position in Europe and appears to not totally forbid "not DTC - based" online genetic testing activities. Since 1997, the European law has regulated genetic testing practices through the protection of human rights and the Council of Europe Oviedo Convention of 1997, particularly through its Additional Protocol concerning Genetic Testing for Health Purposes[1] of 2008, the first international binding legal instrument regulating such practices in Europe. Unfortunately, the Additional Protocol suffered from a lack of

[1] Council of Europe, Additional Protocol to the Convention on Human Rights and Biomedicine, concerning Genetic Testing for Health Purposes, Strasbourg, 27 November, 2008.

binding effect due to few ratifications, France did not ratify this Protocol, even though it is of highest importance for an efficient protection of human rights in cross – border health – related genetic testing for over 820 million citizens. However, in 2011, France ratified the Oviedo Convention that officially entered into force through an application decree[1] and ensured a minimal protection of patients' rights in genetic testing involving laboratories located in a State that also ratified the text[2]. But the specific Additional Protocol goes further into the specific rights of the concerned testees such as informed consent, protection of vulnerable persons, privacy and appears necessary to complete the national legal frameworks in Europe. States always preserve the competence to set up more restrictive policies.

Aside consumer and human rights' issues, DTCGT companies can provide tests which are used within the health systems but also test which are of doubtful quality because only used for research purposes and tests for diseases with no available preventive or treatment measures. The current EU legal framework regarding quality assurance of genetic testing is governed by the Directive 98/79/EC on In Vitro Medical Diagnostic Devices (IVMDD) of 1998. A very important question for the quality of tests and associated results is to know if DTCGT could be considered as using IVMDD or not. This could be clarified in the context of the reform of the EU IVMDD legal framework as the Proposal for a Regulation of 2012, amended by the EU Parliament[3] in 2013, explicitly states its application to predictive genetic tests[4] (in-

[1] Décret n° 2012 – 855 du 5 Juillet, 2012 portant publication de la convention pour la protection des droits de l'homme et de la dignité de l'être humain à l'égard des applications de la biologie et de la médecine : convention sur les droits de l'homme et la biomédecine, signée à Oviedo le 4 Avril, 1997 (1), NOR: MAEJ1221575D, JORF n°0157 du 7 Juillet, 2012 page 11138, texte n° 5.

[2] As planned in the Article L1131 – 2 – 1 para. 3 of the French Public Health Code.

[3] Amendments adopted by the European Parliament on 22 October, 2013 on the proposal for a regulation of the European Parliament and of the Council on in vitro diagnostic medical devices [COM (2012) 0541 – C7 – 0317/2012 – 2012/0267 (COD)], Strasbourg, Tuesday, 22 October, 2013.

[4] See the definition of "In vitro diagnostic medical device" as resulting from the Proposed Regulation COM (2012) 541 final, op. cit. Article 2 (2) and recital (11). Amendments adopted by the European Parliament, 2013, op. cit. Amendment 44.

tegrated in a new risk class C[1], the highest risk class is D), as well as to "device for self-testing[2]" (mainly and surprisingly classed B) including those provided through Internet[3]. Therefore IVMDD made available on the EU market, by European or foreign manufacturers and companies, should meet the general essential safety and performance requirements[4] fixed by the Regulation, and particularly the newly integrated requirements for clinical evidence (clinical utility study and report, usually through clinical trials)[5], in addition to the analytical performance and scientific validity assessments documentation and post-market follow-up. The clinical evidences advanced by the concerned manufacturer are controlled by independent notified bodies in order to obtain the CE marking. The Regulation should also enhance authorities' control and transparency through the updates of the Eudamed database website[6] and the new requirements for better information about the devices "to empower patients and healthcare professionals and all others concerned, to enable them to make informed decisions[7]". It also tends to reinforce the link with ethical principles regarding prior medical prescription, consent, genetic counselling and personal data protection[8]. This new inclusion of DTCGT activities as users of

---

[1] Proposed Regulation COM (2012) 541 final, op. cit. Annex VII, rule 3 and 4.

[2] "Device for self-testing" means any device intended by the manufacturer to be used by lay persons, including testing services offered to lay persons by means of information society services. Proposed Regulation COM (2012) 541 final, op. cit. Article 2 (4) amended in 2013 op. cit. Amendment 46.

[3] Proposed Regulation COM (2012) 541 final, op. cit. Article 5 and recital (15).

[4] Proposed Regulation COM (2012) 541 final, op. cit. Annex II.

[5] Proposed Regulation COM (2012) 541 final, op. cit. Article 4 para. 3; Chapter VII; Annex XII.

[6] Central repository for information exchange between national competent authorities and the European Commission about manufacturers, devices, and market control data. See http://ec.europa.eu/health/medical-devices/market-surveillance-vigilance/eudamed/index_en.htm.

[7] Amendments adopted by the European Parliament, 2013 op. cit. Amendment 14, 15.

[8] E.g. Proposed Regulation COM (2012) 541 final, op. cit. Annex XII, Part A, para. 2.2. Amendments adopted by the European Parliament, 2013 op. cit. Amendment 22 on research ethics committees; Amendment 271 on consent and genetic counselling; Amendment 268 on prescription's requirements.

IVMDD, as validated by the EU Parliament, seems to follow the recent letter[1] from the US federal Food and Drug Agency (FDA) sent in 2013 to Ms. Wojcicki, 23andMe co – founder, that expressly indicated to stop health – related DTCGT activities, called "Personal Genome Service" (PGS), until obtaining the necessary medical device pre – market authorisation requested under the US law. Consequently, 23andMe withdrawn the whole health – related offers from its website and displays specific information about this for consumers[2]. The FDA reaction is a strong sign of law enforcement to this e – market. While it has been very commented as a "crack down" of DTCGT industry historically based in the USA, we wonder if these regulatory activities in Europe and USA will inspire/impact Chinese market.

2. Direct – to – consumer genetic testing and the law in China

2. 1. An activity currently suspended due to the recent genetic testing regulation in China

Before 2014, for the whole industry of genetic testing in China, there was not a law or regulation for gene testing. The providers of genetic testing services through the health institutions or the internet, were all in a state of no supervision.

In February 2014, the first statutory regulation for genetic testing in China has been adopted. The China Food and Drug Administration (CFDA) and the National Health and Family Planning Commission (NHFPC, similar to Ministry of health) jointly issued a regulation, the "Notice on strengthening the clinical use of products

---

[1] FDA, Letter to Ann Wojcicki, 23andMe, Inc. 11/22/13, 22 November 2013. Available at: http: //www. fda. gov/iceci/enforcementactions/warningletters/2013/ucm376296. htm (Accessed on the 1st of September, 2014).

[2] "We no longer offer our health – related genetic reports to new customers to comply with the U. S. Food and Drug Administration's directive to discontinue new consumer access during our regulatory review process". Then, in substance, the website explains that previous consumers keep access to their already generated profile; those who purchased or have purchased 23andMe's Personal Genome Service (PGS) on or after November 22, 2013 (date of compliance letter issued by the FDA) will receive their ancestry information and uninterpreted raw genetic data or will be eligible for a refund. See https: //www. 23andme. com/health/ (Accessed on 4 September, 2014).

and technologies related to genetic testing[1]", that suspends all the genetic testing services in China continent. Indeed, according to this regulation, all technologies, testing instruments, diagnostic reagents and medical software related to the genetics testing shall not be used before having CFDA approval. At that time, none of the genetic testing is permitted, because there is never a licence for genetic testing that has been awarded before, it was no supervision. Therefore, the industry of genetic testing was in a state of shock in China. After one month from the promulgation of the first regulation, the genetic testing services were again allowed under certain specific conditions.

In March 2014, a month after the first genetic testing regulation, the NHFPC issued another regulation, the "Notice on the declaration for pilot units on the clinical application of high – throughput gene sequencing technology[2]" calling for the declaration of pilot units from the providers of genetic testing. This regulation states that any provider of genetic testing, without any difference regarding the institutions or the supply mean used (thus including DTCGT), shall declare its activities and apply for a qualification as "pilot unit" by clearly explaining the types of tests corresponding to their activities and competence. By this way, just like "good examples", some big companies will be selected and authorized to restart the provision of genetic testing services. This is to say that genetic testing, including on the internet, should be suspended for authorisation and that presumably only a small number of big companies who are believed to be technically strongest, with official background will be able to restart. In fact, some companies have never stopped carrying out their testing activities, but they are more cautious.

After the implementation of the two first regulations related to genetic testing,

---

[1] CFDA, "Notice on Strengthening the Clinical Use of Products and Technologies Related to Genetic Testing", http://www.sda.gov.cn/WS01/CL0845/96853.html, 09/02/2014. (Accessed on November 2014).

[2] HFPC Guizhou Province, http://www.gzwst.gov.cn/front/web/showDetail/6220, 14/03/14 (Accessed on November 2014).

the intention of the Chinese medical authority to strengthen the control over genetic testing market and to set up a standard market access system is clear. This should allow eluding the existing confusion regarding genetic testing market and uses in China by monitoring the genetic tests, legally qualified as "medical devices", and the provision of better technical service. According to the adopted laws and the regulatory approach that emerge from them, a new diagnostic technique introduced in China should usually and successively have an experience of "no supervision – pilot unit – standard – rapid development". To date, the genetic testing entered in the second and third parts.

Now the problem is that, except from the two regulations mentioned above, there is no other law/regulation to standardize the market of genetic testing service, whether they are provided onsite or online, what cause particular challenges regarding human rights protection. Therefore, legislative problems in genetics and (online) genetic testing are far from being solved in China.

In China, any genetic testing, including online, should respect the ethical principles fixed in the World Medical Association Declaration of Helsinki of 2013. Indeed, the same challenges and risks than those noticed in France and Europe are at stake in China. They are even more important in a context where the National legal framework is not very developed and where the population lacks from appropriate health and genetic literacy. Despite the new 2014 regulations, the DTCGT offers from Chinese companies remain available through the Internet. In the following section, we will take some case studies in order to describe the reality of the Chinese market and to confront the services provided with the international ethical standards that should be respected to protect customers.

Some known legislative problems in genetics and (online) genetic testing are challenging the Chinese government.

1. Fragile binding force: In China, there is no real law for the gene science, neither for the (online) gene testing. All existing regulations relative the gene science have been promulgated by the State Council, the NHFPC, the Ministry of science and technology, Ministry of agriculture etc. , the legal effect of these regulations is relatively low, they lack of general binding for the whole Chinese society.

At present, we are rapidly developing genetic science and technologies; however, the legislation related to genetics has been seriously lagged behind in China. This is not conducive to the role of law in the maintenance of people's health, of the protection of life security. So it is not difficult to understand why there is a strong call for the establishment of a real basic law in gene such as "Human genetics law" on which build better legal and ethical frameworks for the present and the future.

2. Lack of systematicness in the legislation related to health genetics: In China, there are a dozen regulations concerning genetics. Existing ones include activities of biological and agricultural research, genetic reproduction, gene therapy, research on embryonic stem cells and research on cloning. These regulations have been promulgated by different departments of the government, and every regulation has been formulated for one detailed target. There is no clear or only a small correlation between these regulations. In addition, due to the lack of a Chinese basic law in genetics, it is quite long to formulate, adopt and coordinate secondary laws/regulations. As a result, there is a lack of consistency among existing regulations. Some important areas of human genetics technologies are not yet properly regulated and other crucial domains remain regulatory lacking such as privacy protection. Existing regulations and new ones will need a unity concept to guide the development of gene science and technology in harmony with the full respect of human rights.

At the same time, the cohesion between these regulations and the other legal departments is not satisfying, because the legislative branch does not formulate and promulgate the necessary regulations to coordinate between them. Therefore, behind the basic legal works to undertake in a near future there will be another important work to perform on articulating different laws.

Considering this, the constitution of a legal system on genetics including is absolutely necessary.

3. Lack of theoretical study and actual application: In contemporary age, the development of genetic science and technologies is important to cope existing and future health challenges. The progress in genetic research will take a more and more significant role in our life. This will also produce a large number of new social relationships where some new rights and interests will be challenged, such as in the domain of gene patenting. However, today, existing studies of the Chinese situation are insufficient to debate about the model that would be fitted to China. More national studies should be performed in China. Various foreign models such as in the EU, the USA and the Canada should be studied in order to figure out the best regulatory approach for China. The lack of legislation in human genetics reflects this reality.

Moreover, for some, the affirmative opinions regarding the protection of privacy or informed consent in genetic testing through existing regulations on genetics suffer from a too theoretical approach. In Chinese laws, these concepts and relative rights and obligations are not clearly defined. Therefore, they are hard to understand and to implement in practice. Such principles of human rights in biomedicine are thus widely lacking from essential elements of enforcement. This includes, for example, the fact that there are no clear penalties for the possible violations.

2.2. The Chinese DTCGT markets' reality

In recent years, the Genetic testing industry is developing rapidly in China. There are more than 200 genetic testing companies and more than 120 independents laboratories that obtained the approval from the China Food and Drug Administration (CFDA) and that mainly provide non-invasive prenatal and tumours genetic testing services. Such a situation is due to the following reasons. The disadvantages of the traditional detection methods are increasingly obvious with the development of new genetic testing technologies[1] whose cost has been greatly reduced. Now the cost is

---

[1] For example, for the prenatal foetal birth defects inspection, traditional amniotic fluid puncture method has some shortcomings as the high risk and low accuracy, and new type of prenatal genetic testing is safe, non-invasive, accurate, and zero risk.

only 3000 RMB in China (1 USD = 6. 17 RMB). And in the near future, this price will be reduced by half. The health needs in China orient towards the use of genetic testing technologies notably prenatal genetic testing, as China has a high incidence of birth defects[1] and the simple, secure, accurate, no risk genctic testing practice can discover that the foetus is at risk to suffer from diseases. The existing supervisory system does not essentially constraints the development and promotion of this technology. From policy level, Chinese government has even been encouraging the development of genetic testing industry, before 2014 there was not too much state intervention to limit the development of this industry.

At present, in China, the most implemented genetic testing is the non - invasive prenatal genetic testing mentioned earlier. This market has mainly been driven by the two Chinese companies (BGI and Berry Genomics, the former finished 100 000 cases of genetic testing and the latter finished 50 000 in 2013), that worth about 450 million RMB each year and the overall market of genetic testing worth 1 billion RMB. In addition, because the costs of non - invasive prenatal genetic testing provided by Chinese companies are far below than the abroad price, the quarter customers of the two companies originate from abroad. BGI has accumulated about 300 000 prenatal screening of genetic testing for down's syndrome, that occupy half of the global total, around 30% of them are from overseas.

In China, many public genetic testing institutions open the online consulting and/or offer the tests, while nearly all private institutions offer online services, even some offer specially the genetic testing services. Then with the latter as the example shows the process of implementation for one online genetic testing. If a customer interested is by a test in site, after his registration online, he can consult the agency staff for a communication about the choice of genetic testing, with the professional

[1] Currently about 5. 6% (5% in 2002), every year there are about 900 000 cases of birth defects in babies, among them about 250 000 clinical visible birth defects.

guidance, the customer can make a decision on genetic testing and make a payment for the corresponding costs online. Of course, this customer also can choose the test (s) and pay directly online without the consulting. Then, based on the actual needs of the test, the customers can be divided into two groups. One group needs to go to the health establishments for the collection of the sample for testing; the other group of people will receive a sample collection device, complete the collection and package by themselves, and send this device back to the company. About the inspection report, all customers are able to use their individual mailbox to query the result on the site. Also, they can consult the professionals of the company in order to make clear the practical significance of the detection.

There are some examples explaining the process of online genetic testing in China. We will now focus on the services of two Chinese most well – known companies proposing DTCGT, for breast cancer predisposition detection for example.

Example 1. Berrygenomics (http: //www. berrygenomics. com/En/Default. aspx), located in Beijing, who is the No. 2 in the Chinese market of genetic testing. In its official homepage, there is a simple list of their testing services, mainly for the non – invasive prenatal testing. If somebody needs some more detailed information, he/she should call their customer service phone. Then, after a telephone consultation, their professional advisory staff express clearly that they cannot provide for the breast cancer testing services. They do this testing just for scientific research and do not provide products for the public. Thus, we are more here in a direct – to – professionals approach than in a direct – to – consumer one.

Example 2. BGI DX (http: //www. bgidx. cn/index), the subsidiary of BGI (http: //www. genomics. cn/en/index) who is the strongest genetic testing company in China continent, it is also the No. 1 in the world testing market. At the same time it is the institution in gene scientific research that is the most supported financially by the Chinese government. BGI is located in the city of Shenzhen, founded in September 9, 1999, BGI DX, professional engaged in genetic testing service to the public, was founded in July 2010. BGI has the world's most advanced genetic testing instrument and equipment, in the market of the non – invasive prenatal genetic testing, it take more than 50% share. To purchase a service of breast cancer genetic testing, the person must enter the official homepagc of BGI DX (http://www.bgidx.cn/index) and click on the "Services items" that will display all the names and types of testing products. In our example, the person click on their breast cancer testing (because the breast and ovarian cancer is associated closely in gene BRCA1/2, so the two tests of these two cancers are combined together and did at this time). The basic information about this test are displayed, e.g. the background of testing, the service content and clinical significance, the technical advantage, the target population, the service process, requirements to sample, typical case, common problems, etc.. In the understanding of the above content or not, if the person is really interested, she can call the customer service telephone (400 - 605 - 6655) or by social software (QQ, Chinese most widely used in social software), for more information and consultation. In the process of communication, the staff will register some basic information (specially the choice of testing), and strongly recommend that the consultants need to have a specific consultation with a doctor and make a correct decision for genetic testing. If the person really have intention to have a deeper understanding of some genetic testing, the company staff will once again confirm and register the surname, telephone number and the main motivations of testing, as well as for example, personal and family medical history etc.. Nothing is mentioned about data protection on the public website. Then, the company will constitute a medical professional staff, a doctor perhaps, who is responsible for a deeper and more concrete consultation with the person, just like the confirmation of testing type and the price. After such consultation, the person then does a final decision on whether to buy the test.

Once the people has finally determined to do the testing for breast cancer, then the customer must first pay, then BGI DX will arrange the customer to the designated hospital to collect blood for testing. Because the majority of customers are not in the Shenzhen city, BGI DX has cooperative medical institutions, normally hospitals where the customer will be received for the consultation and collection of blood, in major cities of the country. In fact, there is only a limited number of doctors who really and correctly understand the result of genetic testing, so BGI DX has certified some doctors in theirs cooperative hospitals. These doctors are mainly responsible to explain the result of testing and give their medical recommendations to the customer.

After the analysis, BGI DX will send testing results to customer with the medical significance of results in detail. If necessary, the customer could call again or contact the company staff through the network in order to understand the results of testing. Of course, the customer could go to the cooperative hospital to see the doctor. In one word, it is the customer who takes the decision about how to understand the testing result. In general, the breast cancer testing period is 35 working days.

Through the above examples it can be seen that DTCGT in China mainly has following several problems:

1. The offer of DTCGT is discontinuous and unstable experienced a course "no regulation → completely ban → a few pilot unit". Also, present authorized type of DTCGT is quite limiting and cannot meet the actual needs of the market.

2. Through the investigation of the author, at present, DTCGT even in the pilot unit company, informed consent for testing is incomplete or imprecise, and even in many cases it has been intentionally or unintentionally ignored. The company just informs about the medical significance of testing.

3. In the view of supervision for DTCGT, there are still many loopholes. Although only a handful of companies are authorized to carry out this service to the public with a few types of testing, on the Internet, it still is easy to connect to some unauthorized company to carry out various types of testing, even invalid, or medically useless.

Discussion

DTCGT is a quite recent phenomenon deserving regulatory considerations in the light of the benefits and the risks that such a commercial approach of genetic testing entails for the person and for the health system. Several regulatory models could be used to appropriately frame these activities in China. In France, the prohibition seems to be based on the imbalance of risks and benefits of DTCGT. The problem is to ensure enforceability of this principle and to ensure sufficient capacities of traditional health system in terms of only prescribed evidence - based medical genetic testing. Restricting the access to DTCGT is an option but it is restricting individual freedoms and obliges the State to provide all the necessary health genetic testing services to the patients in order not to deprive the people from care or prevention means. Furthermore, it does not prevent other direct - to - professionals (prescribers, researchers) offers of genetic testing services performed by private companies. In China, while there is no specific legislation on DTCGT and only few about human genetics, the DTCGT offers are presumably legal. Since 2014, Chinese government adopted a market control - based approach by obliging the assessment and registering of any genetic tests by the CFDA through two new regulations. The first one has shocked the market, and the second one recovered it by strengthening a national oversight system of medical devices such as genetic tests. However, the number of companies has been strictly limited to those obtaining the license. This kind of management may be a bit simple and rude. It could especially harm the interests of consumers and small or medium companies. Indeed, the current needs of patients cannot be fully satisfied by traditional health system, hence the attractiveness of a DTCGT model for China that would be enshrined within existing health system; provided that medical genetics skills be improved within hospitals. Economically, while small company's business activity has been suspended, some of them have to continue their activity without authorization in order to survive. Despite this economical effect, the current trend pushing from "no supervision" to "supervision" reflects the government's consciousness of its supervision responsibility and must be saluted as it should gradually improve and standardize genetic testing activities. While it is essential to protect citizens and avoid the misuses of human genetic resources, like excessive privatization, a great part of the ethical and legal issues regarding human rights in biomedical activities, biobanks commercialization, privacy protection and medical ongoing of the consumers remain insufficiently regulated or controlled. The sole control of the testing products will not suffice to ensure appropriate, safe and useful genetic health services. After a period of rectifying and standardizing the market regulation and oversight, we can guess that DTCGT could meet their expectations and respect the rights for the maintenance of health without completely banning it as some foreign countries did.

THE RIGHT ATTRIBUTE OF THE GENETIC INFORMATION: A KIND OF COMPOUND INTEREST

CHEN Zihan [1]

People used to take themselves as a dividing line of interests in less developing era of natural sciences, so the relationship between man and the world is defined as "people" and "object". However, new situations and new problems that are brought by the development of modern science and technology is guiding the theory of rights and transcending the construction of two between "people" and "object". This trend is reflected from "intellectual property rights" becoming the third kind of legal interest apart from "personality" and "property" in the field of the traditional private law to the protection of compound interest advocated by the right attribute of the genetic information.

1. The Phenomenon of Rights above the Right Attribute

The connotation of gene information has two core elements: Information and obtainment of special analysis method. Based on this definition, reflection on the particularity of genetic information of objective and subjective aspects, is the basic cognition on the genetic information as the object of right. Meanwhile, it can also affect the right attribute. Based on the particularity of genetic information, individuals have access to genetic information demand. In family relations, people have demands that blood bond realized the health state of individuals by obtaining others' gene information and spouse try to know fertility gene and the risk of life by getting other party's genetic information. But in the field of education, employment, insurance which is

[1] Lecturer in law school of Beijing Institute of Technology, Doctor in law (Shandong University), PhD in Biological and Medical Law (University Pawl Sabatier).

closely related to personal status, people think genetic information is related to the cost and risk of their own . The scientific research institutions have more intuitive benefit demand on the gene information, picking a fight with the providers of the gene information about "property", from the aspect of public interest, set up the database to collect and process a large number of the genetic information for medical research, administration, judicial, military security and so on. We analyze the adjustment of exclusive power to hold the genetic information, privacy right to the circulation of gene information, equal power allocation of resources related to the genetic information, is for the analysis of property rights reasonably, excavating the point of compound interest behind the phenomenon of right.

1. 1. The Holding of Genetic Information

Nowadays, the consideration of issue on balancing the ownership of genetic information is no longer confined to the problem whether it is being patent by eliminating the valuable finding from public interest, and the issue on balancing the interest behind the genetic information should be paid more attention. In that moment, the subject of study and submission of biological information is not popular, which is still confined to the study institution and the lab of Drug Company. Unfortunately, it will take a long time to be diagnosed and analyzed by doctors directly through patients' genetic information, which is one of the most important goals of genetic information study on biological medical science.

The latest news about genetic information has aroused public attention. One is "Hela Cells Argument"[1] published in Nature. Hela Cell, collected before more than half century, transferred the fruit that pushing the technical progress in the field of biological medicine but under the condition of unawareness of Lacks' Family. The decision of Lacks' Family after half century is another unselfish contribution unques-

[1] "Nature: Hela Cells Dispute Has been Satisfactorily Resolved", contained Web Biological Exploration, http: //www. biodiscover. com/news/research/105563. html (last visiting date: 2015. 03. 01).

tionably. The other is the Phoenix reported a story quoted by RIA Novosti in Nov. 27th, 2014, said "the U. S. Food and Drug Administration, FDA, was banned on 23andMe, Inc. in offering personal genetic detected service with no certificate". [1] The U. S. 23andMe, Inc. promoted the genetic detected service in a low price since 2007 with the aim of drawing the consumers to purchase the service that is to gain one's personal statistic analysis of Genome by mailing the sample of saliva, and the price of the service was finally dropped into 90 U. S. D with the original of 299 U. S. D. FDA was not against the behavior by U. S. 23andMe, Inc that attracts people to realize the knowledge of inheritance and gene in object of themselves by U. S. 23andMe, Inc..

While the company began to put their focus on providing premonition, relief and treatment to the patients who is suffering from cancer, metabolic disease or in healthy risk, and sale the medical apparatus and instruments produced by the company, and this operating activities must pass the FDA's permission. The selling were shut down. However the real trouble is not the selling without FDA's permission but 23andMe is designed to get amounts of information of personal gene data by selling their service of checking at a low price. 23andMe has became a way that is unknown to the public to collect gene data, and the gene data that they collected is likely to be sold to the insurers or the business institutions which in the requirement of gene data. [2]

The two instances indicate vividly that scientific research institutions and medicinal companies will exploit, invest and contend for the gene source, gene information and gene knowledge driven by the capital benefit. Their researches of gene have

[1] "U. S prohibited the Company 23 and Me without A license to Provide the DNA Testing Service", contained Web Phoenix network, http://innovation.ifeng.com/biz/detail_2013_12/06/1557224_0.shtml (last visiting date: 2015. 03. 01).

[2] Charles Seife: "23 and Me Is Terrifying, but Not for the Reasons the FDA Thinks", Scientific American Web http://www.scientificamerican.com/article.cfm?id=23andme-is-terrifying-but-not-for-reasons-fda (last visiting date: 2015. 03. 01).

some features: restoring gene samples infinitely, using gene samples whenever necessary, limiting the sharing of data or samples by taking advantage of the stability of DNA, making a strict control over the access, using and publication of the gene information, limiting gene recycle, and paying attention to the connection between families and communities. These characteristics of gene information research institution decide the tension of information supplier and institution itself. The research institutions want to develop secretly, to build the data base of their biologic business, to get simple permission of gene and to get the preemption of the research of the gene information.

Relatively, the suppliers of gene resource, information and knowledge are asking them to share the benefits of gene business to against the "gene pillager" or "gene pirate". Information suppliers should make a specific agreement that the institutions can't put human bodies into business, suppliers have rights to access to samples, and institutions have to take care of the samples as well as they can. The institutions also should promise that the gene data and samples will be more open and sharing.

Above all, a duty behind the question whether genetic information is property and disputes about intellectual property, is the establishment of ownership and the standard of use, which contains rights of control, announcement, management, secrecy more than the ownership. In summary, the core powers and functions for the providers of genetic information are to know the state of genetic information and how the information being handled. Otherwise, for the users of genetic information is the priority. The purpose of the priority of the users of genetic information is to balance their scientific research investment, which may encourage technical progresses. However, considering the characteristic of genetic information, the time limit and the fair use of this priority, which is the key to our right, needs scientific and reasonable practice in legislative activities.

Meanwhile, to the providers, our right still contains the informed consent, in which the "informed" matches the informing of the researchers and the "consent" means independent and free choice made by the providers. Of course, it's not the first time the informed consent used but rather accepted as a fundamental principle by people when it comes to human biology and medicine. However, depending on the danger which providers face and the aim, such as the gene research, gene therapy or the genetic information database for administration, the informed consent must change itself through the use of genetic information. The informed consent must clear the following: the research doesn't end, and the informed consent doesn't either; the content of the informed consent contains uncertainty; privacy risks and uncertainties may exist for a long time; the informed consent is a kind of limit to the independent and free right of providers; the providers have a right to withdraw their consent. Giving the priority to the researchers and the informed consent to the providers, the law can ensure that the public interests and the scientific research are able to benefit from it.

1. 2. The Transmission of Genetic Information

Where exclusive right of genetic information may change the condition of possession and use of genetic information through priority and principle of informed consent, the power and function to privacy is the transmission of genetic information. As mentioned above, regarding to the rational choice on how to protect the genetic information, it is necessary to base on the objective knowledge of particularity of genetic information. The relevant dispute about the right of privacy of genetic information is exactly the specific standpoint where economic filed treat the particularity of genetic information.

Since the Warren and Brandeis published the Right of Privacy on Harvard Law Review in 1890, until now, the theory and system, as well as the practice of legal system of the privacy right have been further developed. The right of privacy consists of the body privacy, space privacy and information privacy[1], that is, a control of personal privacy filed. The scope of protection of privacy is limited to the personal information in civil law system and especially below the China current legal system[2]. The purpose of privacy right is to guarantee individual autonomy, and further guarantee the capacity of safeguarding self - interest in social interactions. The individual obtains the resource superiority in social life by means of providing and preserving the personal information.

The genetic information is distinct from the general identity information, and it can be obtained by virtue of given detection means. So the right of privacy of genetic information first shows the individual control of information, that is, the right to access genetic information. In practice, it is expressed as the right of requiring the doctor to notify the gene detection result and to consult relevant genetic information data at genetic databases. Declaration of the International Human Gene Data Article 13 defines the individual is entitled to inquire and acquire the gene information[3], and that the clearly know the situation of genetic information is the premise of protecting the right of privacy.

On the other hand, individuals have the right to choose not to know his own genetic information, which is also protected by genetic information rights. James Watson, honored as "DNA Father", is one of the co - discoverers of the double helix structure of DNA. After 50 years of his discovery, Watson provided his own blood

---

[1] Julia C. Inness. Privacy, Intimacy, and Isolation, Oxford University Press, 1992, p. 95.

[2] Although in the civil law system, privacy right takes attention of the self - determination of private affairs, in the traditional theory, property right is not included in the protection systems of privacy right. With existing of the right of life, the right of health, freedom, the scope of protection of privacy is limited only to the protection of personal information data. Moreover, whether it is in Chinese "Constitution", "criminal law" or administrative law, civil law and the relevant procedural law norms, provisions about privacy are limited to the level of protection of personal information.

[3] UNESCO, International Declaration on Human Genetic Data, Article 13 - Access, No one should be denied access to his or her own genetic data or proteomic data unless such data are irretrievably unlinked to that person as the identifiable source or unless domestic law limits such access in the interest of public health, public order or national security.

sample to 454 Life Sciences Corporation which works on gene testing and his genome was mapped in one month by the company. However, Watson and 454 came to an a-greement demanding the company not to reveal all his DNA information to himself. [1] Because the Watsons have a heredity disease of Alzheimer's, i. c. senile dementia, therefore he isn't desirous to know whether his APOE gene concerning about the dis-ease has varied or not. Article 10 under the International Declaration on Human Ge-netic Data empowers people who received the test to determine whether to be in-formed of the research results or not [2]. Article 5 (c) under UNESCO's Universal Declaration on the Human Genome and Human Rights also underlines that "The right of each individual to decide whether or not to be informed of the results of ge-netic examination and the resulting consequences should be respected" [3].

Therefore, individuals have a right to decide whether to know their own genetic information or not, which is the premise to achieve the self-ruling of genetic infor-mation. Besides, genetic information right can also guarantee individuals' interest of keeping genetic information in privacy. This kind of interest can be reflected in the collecting, utilizing and holding of their own genetic information. In this regard, In-ternational Declaration on Human Genetic Data clearly stipulates the collection, use, storage and circulation of "human genetic data, human proteomic data and biological samples". Thus the right of privacy ensures that individuals providing his genetic in-

---

[1] "The Genome of Father of DNA Will Be Decrypt", contained Xinhua Website http://news.xinhuanet.com/world/2005-05/28/content_3013288.htm (last visiting date: 2015. 03. 01).

[2] UNESCO, International Declaration on Human Genetic Data, Article 10 - The right to decide whether or not to be informed about research results: When human genetic data, human proteomic data or biolog-ical samples are collected for medical and scientific research purposes, the information provided at the time of consent should indicate that the person concerned has the right to decide whether or not to be informed of the re-sults. This does not apply to research on data irretrievably unlinked to identifiable persons or to data that do not lead to individual findings concerning the persons who have participated in such a research. Where appropriate, the right not to be informed should be extended to identified relatives who may be affected by the results.

[3] UNESCO, Universal Declaration on the Human Genome and Human Rights, Article 5 (c) - The right of each individual to decide whether or not to be informed of the results of genetic examination and the resul-ting consequences should be respected.

formation has the right to control his owns. The right of privacy manages the genetic information to realize its circulation.

However, the genetic information rights ensure the achievement of individual social participation and the origin is not only the individual value, but also should consider the social impact brought by the achievement of the rights. For instance in the medical field there are needs that the genetic family association of information makes the members of family have the demand to desire to get others' genetic information to know the sibship and inference the health condition of themselves. Especially in the relationship between treatment, the requirement of the family members to the genetic information turns into the doctor's obligation of disclosing, and this is the conflict between the right of privacy and the right to learn the truth of the genetic information where the measurement of it is between the genetic information's hidden value and the health of the family members.

Of course, there is also a problem in the field of insurance: whether the insurance company have the power to ask the policyholder to provide the genetic information, and if the policyholder could check the genetic information when his right of privacy is violated or when they make the check of genetic. For one hand, we should consider the particularity of the genetic information which could ensure if the genetic information in the field of insurance distinguish other information involving the individual privacy. In general, the information you are demanded for when you go to take a insurance, such as the age, sex, the illness of family history which are all associated with the estimation to your health, and it also infer to your family and your association, getting the genetic information won't make the insurance company make any profit overcoming other personal healthy information. The purpose of insurance is to share the risk, in which the state of fair is the person who has the same probability of risk to invest the same proportion. The uniqueness of genetic information could insure the people who carry the defective genetic and the malgenic genetic based on the high risk and pay more than the normal. If we resist this kind of divided risks for the privacy right, it will lead to the unfair situation in the divided risks, and make the field of insurance's spontaneity and rationality in the distribution of resources.

On the other hand, we have to consider and weigh the social influence on the genetic application in the field of insurance. The Life insurance has become an important life safeguard for natural person in the society with rule of law. In some countries where social insurance could suffer the basic personal risk, it provides individuals with equal protection. The individual's right to survival and health will not be under threat because commercial insurance will not propose higher premiums on the people carrying pathogenic and defects. In some countries and regions where commercial insurance share the risk with the most people, insurance threshold will increase the burden of the people in enjoying basic rights. It is so called the selection between the interests of insurance company property and the benefits of the individual life and health.

In the fields of employment and education, employers and schools tend to obtain genetic information and avoid the risk with the reduction of the genetic information acquisition costs. Therefore, the right of privacy controls the circulation of genetic information, which aims at balancing the individual value of information and the altruistic value and social value of information.

1. 3. The Resource Configuration of Genetic Information

Declaration of the international human genome data first defines the equal principle about the acquisition, processing, use and conservation of genetic information.[1] However, inequality is still widespread due to development of genetic technology. This inequality includes two kinds of conditions, and the first cannot use equally the genetic information due to unequal economic resource, such as gene therapy and gene detection. This equality doesn't exist only in the resource configuration, but also in the every aspect of economical life, and it can make more people enjoy the benefits from development of technology from the aspect of economical development, increase of income and decrease of technology costs. In other situations, inequality shows up as the genetic prejudice. For instance, according to genetic testing, the cases in which people carrying "Huntington's carpal tunnel syndrome" genes were fired have already existed[2]. And the case that happened to public official recruit does not employ the people carrying "the Mediterranean anemia" genes has attracted widespread attention. This genetic prejudice aims inequality on people depending on the difference of genetic structure.[3]

The equal rights protection of the power of genetic information contains two connotations, individual equality of opportunity and equality of treatment, the former refers to disregard, while the latter emphasizes the allocation of resources with no

---

[1] UNESCO, International Declaration on Human Genetic Data, Article 1 – Aims and scope.

(a) The aims of this Declaration are: to ensure the respect of human dignity and protection of human rights and fundamental freedoms in the collection, processing, use and storage of human genetic data, human proteomic data and of the biological samples from which they are derived, referred to hereinafter as "biological samples", in keeping with the requirements of equality, justice and solidarity, while giving due consideration to freedom of thought and expression, including freedom of research; to set out the principles which should guide States in the formulation of their legislation and their policies on these issues; and to form the basis for guidelines of good practices in these areas for the institutions and individuals concerned.

(b) Any collection, processing, use and storage of human genetic data, human proteomic data and biological samples shall be consistent with the international law of human rights.

(c) The provisions of this Declaration apply to the collection, processing, use and storage of human genetic data, human proteomic data and biological samples, except in the investigation, detection and prosecution of criminal offences and in parentage testing that are subject to domestic law that is consistent with the international law of human rights.

[2] United States' Department labor, et al, 1998.

[3] Natowicz, R. Marvin, Jane K Alper, and Joseph S Alper: "Genetic Discrimination and the Law", *American Journal of Human Genetics*, 1992.

difference. The equal opportunity not to be ignored, as can be found in 1776 The Declaration of Independence, 1789 French Universal Declaration of human rights, 1918 Russian Soviet Federative Socialist Republic constitution, the UN Universal Declaration of human rights, and China's current constitution, is treated fairly, as well as the basic requirement of the rule of law to protect all social members. It is the proper meaning of fairness and justice that humans should not suffer injustice because of age, gender, ethnicity, disease or disability, including of course the difference of genetic information. As it is in a certain historical period of factors such as gender and racial discrimination, the public understanding of the particularity of genetic information is still on the deviation, to the myth of gene determinism or gene specialism. Blind worship of genetic information leads to the ignorance that the most genetic diseases is a polygenic disease and individual comprehensive factors associated with the environment. Universal Declaration on the Human Genome and Human Rights, Article 6 highlights that anyone should not be discriminated against because of genetic traits. Equal rights conferred by the dignity of personality and the spirit of freedom shall be inviolable [1]. Therefore, different and virulence gene structure and defect gene carriers should be respected in the dignity of human nature, enjoying equal opportunities and possibilities.

On the other side, equally legal capacity of genetic information rights prepares judgment standards for the distribution of sources, which is known as unfair treatment in the real life, in order to judge its reasonability. Realistic rights equality has relativity and rights of equality must be enjoyed respecting the distribution rules in diverse fields of society. In other words, the requirements from fields of society must be considered. Discrimination does not mean differential treatment. Eliminating the discrimination against gene does not mean that we should not distribute sources rea-

[1] UNESCO, Universal Declaration on the Human Genome and Human Rights, Article 6 – No one shall be subjected to discrimination based on genetic characteristics that is intended to infringe or has the effect of infringing human rights, fundamental freedoms and human dignity.

sonably according to the genetic information. Although the skeleton of Human Rights Act system compose rights of equality and rights from discrimination, if differential treatment is reasonable and acceptable objectively in some sense, it tends to be permitted. [1] For example, it exists as a tradition that employers like to collect health information of his employees aiming to stabilize the production. Because it can benefit personal and property security of employees, genetic information rights will not be limited when employers have to choose the best employee according to his health condition.

When genetic information is used to choose applicants by bosses, its particularity should be taken into account. Firstly, it must be clarified whether bad genetic information like normal health information signifies that the employees' performance will be influenced because of, for example, the decease and shortage brought by bad gene. Secondly, genetic information plays a particular role in the choice of law value. Some sensitive genetic information, which can be found in law, must be utilized as prescribed by law and with the permission of people involved. British Bioethics Committee has ever suggested that when employees and third part would be at risk because of disease and particular job and risk and disease can not be avoided by employers, they can obtain genetic information to examine applicants from gene test. Therefore, collection, dispose, use and conservation of genetic information must obey principle of equality. According to the rights of equality, individuals must not suffer discrimination and overlook. And personal dignity must be respected. Meanwhile, the rights of equality respect distribution principles in diverse fields of society and judge their reasonability.

2. Interest Direction of Right Attributes

The purpose of analyzing the attributes of right doesn't merely lie in the con-

---

[1] Zhang Mingan, *Privacy Infringement of Discloses Private Affairs of Others: Medical Information, Genetic Information, Employee Information, Airline Passenger Information and Online Privacy Infringement*, Zhongshan University Press: China, 2012, p. 357.

struction of a specific right or a bundle of rights of genetic information, to find the law to regulate the interest among information users. It is of great significance for a scientific, rational and effective protection frame to analyze the attributes of completed right objects and plural subjects. It has been proved that the establishment of genetic information right system is highly effective through the legislation practice of USA, EU and other international organizations. But whether the rights bundle could embrace all the interests behind genetic needs to be proved by combing the interest disputes of it, which is the most important part in the system construction.

In China's civil law, the cognition of personal information interest, lies in the origin of tradition which is theoretically based on the work of Warren and Brandeis, "The Right to Privacy" in Harvard Law Review. [1] The representative point in the field of civil law is of the Professor Zhang Xinbao in 1997, "the legal protection of privacy", Professor Yang Lixin in the 2006 "type of tort law research", Professor Zhu Yan in 2011, "tort liability law" theory in the making of this statement; in the civil law field, Wang Liming and Professor Yang Lixin also uphold this in 1997 in his "personality right law" and the related elaboration of personality right law; in the civil law field on the mainland, our far - reaching Taiwan scholar Wang Zejian, also believes that the founder of the theory of privacy is USA scholar Warren and Brandeis; in addition, in the field of protection of the rights of genetic information, in Luo Shenghua "the protection of gene privacy" the origin of the theory of right of privacy is considered to be the United States. At the same time, legislators in the legal working committee of the Standing Committee of the National People's Congress gave an indication of the tort liability law background and ideas, believing that the right of privacy theory originated from the United States. So we used to take Warren and Brandeis as the raiser of the problem.

---

[1] Samuel D. Warren and Louis D. Brandeis, "The Right to Privacy", *Harvard Law Review*, Vol. 4 (1890), pp. 193 - 220.

Professor Zhang Mina, his book "the comparative study of the right of privacy—France, Germany, the United States and other countries of the right to privacy" [1] summarizes our reason to recognize "the right to privacy" as the originator of the theory of right of privacy, mainly influenced by Taiwan scholar Lv Guang, "the mass media and the law" [2] and "the legal value of western society" published in 1990 by Peter Stein and John Shand [3]. The scholars in our country have a broader interpretation of Professor Lv Guang "the mass communication and the law" that "the theory of the right of privacy in the common law didn't get widely recognized before the publishing of 'the right to privacy'", the meaning of the right to privacy for the common law has extended to the common law as well as other countries.

And in the book "western society in the legal value", because the protection of the right of privacy laws just be written into the "French Civil Code" Article 9 in 1970, the citation of the law as trial practice experience is rarely expressed thus caused misunderstanding that no right to privacy and related trial practice before. [4] Zhang Mina professor believes that the French civil code since 1970 in the supplementary article 9 began regulation about the right of privacy of the academic expression is deeply influenced by "the western society in the legal value", a dramatic illustration of this is that the translator of "the western society of law" make supplement the French Civil Code Article 9 date by July 17, 1970 mistaken in translation as of July 11, 1970, and thereafter the civil law learn basic writings also has been in use as a new French Civil Code Article 9 birth date on July 11, 1970. [5]

The French Civil Code of 1804 did not provide "the right to privacy", but that

---

[1] Zhang Mina, *The Comparative Study of the Right of Privacy—France, Germany, the United States and Other Countries of the Right to Privacy*, Zhongshan University Press: China, 2013, pp. 16 – 19.

[2] Lv Guang, *Mass Media and the Law*, The Commercial Press: Taiwan, China, 1982, pp. 65 – 66.

[3] See Peter Stein and John Shand, *The Legal Value of Western Society*, translator Wang Xianping, Chinese People's Public Security University Press, 1990, pp. 224 – 250.

[4] See Zhang Xinbao, *Legal Protection on Privacy Right*, Masses Press: China, 1997, p. 50.

[5] Zhang Mina, *The Comparative Study of the Right of Privacy—France, Germany, the United States and Other Countries of the Right to Privacy*, Zhongshan University Press: China, 2013, p. 26.

doesn't mean the French civil law system can not protect genetic privacy interests when it is infringed. In 1804 the French Civil Code article 1382 and 1383, the provisions include that personality and property interest violations shall bear the responsibility. [1]

There is no denying the fact that Warren and Brandeis on the privacy of the information development has made outstanding contributions, but the meritorious behind also concealed the other way to protect the information. Our country, since the end of the 20th century, took the genetic information rights and all types of information rights of the Warren and Brandeis academic achievements as the origin that over a long period of time, Warren and Brandeis positioning theory and the the oretical basis also largely limits the trend of information to protect the interests. When facing the new problems of the protection of genetic information, this is also one of the reason of the dilemma. Thus the origin of right attribute of information, not only lies in returning to the history, the more important significance lies in making the colorful right protection system facing the reality and the future, finding and solving problems.

2.1. A reasonable expectation of the interest of the genetic information

In American, Warren and Brandeis are not the people who created this privacy protection tool but made detailed information related to the tradition of the protection of the interests and the legal norms summarized and abstracted. And the American has the judicial tradition with respect to individual freedom and privacy, which makes Warren and Brandeis concluded based on "the right to independence".

And Warren and Brandeis also absorbed the continental law system in protecting the private living space, with an emphasis on the protection of personal emotional reasons. Warren and Brandeis' creation of "the right to privacy" is in an era that

[1] See Zhang Mina, *The Comparative Study of the Right of Privacy——France, Germany, the United States and Other Countries of the Right to Privacy*, Zhongshan University Press: China, 2013, p. 27.

the U. S. news, media and publishing industry was booming and expanding, new technology and communication tools were widely applied into people's daily life, in addition to the protection of self space, they think privacy lies in the protection of the individual personality and impaired emotion. However, under the protection of the individual freedom, Warren and Brandeis' awareness of the protection for the interests of the individual, is closer to the in a "world power".

They keep information privacy violations defined under the situation of the stranger's access and disclosure of personal information although actual condition is much wider and the era of information protection theory is also being enriched, there is no doubt that guided by Warren and Brandeis, the privacy development path is built on personal freedom and the protection of personal feelings. However, Warren and Brandeis' theory of the right to privacy definitely not only affects the development of the privacy tort law, in the second year published in the "right to privacy", "right to independence" was included in the constitution and has withstood the study section of the first amendment to the constitution of the United States. Although there have been a case that because court had the access to information in a legal way and the victim's name were released without punishment, this principle was eventually overthrown by the legislation and judicial practice. Because except for academic research breakthroughs, what is more important of the protection of rights is to obtain the effect and support of the people. It has been a profound personal self – conscious.

William Prosser is the scholar who further stylized and immobilized the tort law protection mode of the right to privacy. Prosser summarized at all levels of the United States Court about 300 cases since the article of "the right to privacy" was published to the middle of the 20th century. He believes that these cases to protect the interests can be summarized as four types: first, intrusion into other person's residence or disturb other private affairs; second, making others embarrassing private facts public; third, publicly vilifying others image; fourth, unauthorized use of the person's name, portrait.

Subsequently, Prosser took this view into his "the American tort law Restatement (Second Edition)" with the reporter of himself, which laid the foundation of the United States Information System of tort law with Warren and Brandeis theory as a starting point and Prosser stylization as the milestones. So before the United States enacted special legislation on the protection of the gene information, one of the important basis for the protection of genetic information is as the object of the protection of tort law.

But as previously have repeatedly emphasized, the significance of the times of the Warren, Brandeis and Prosser is not great enough to summarize genetic information related to the protection of the interests of all. They were neither unprecedented, nor exhaustive. This investigation tasks are mainly based on two points. First, Warren and Brandeis and Prosser ignored that before the publication of "the right to privacy", the United States can protect part of the theoretical basis of the rights of the gene information; second, as the contemporaries of the USA, British and continental law countries, studied the personal information protection in another way, which also is of far-reaching significance to explore genetic information rights protection today.

In the United States, before the stylization of personal life including personal information tort, the protection was mainly based on two theoretical background. The first is that Warren and Brandeis advocated the concept of the right to privacy, which is mainly based on the inviolability of the personality. Their main basis is the famous British case Prince Albert v. Strange[1] and the discussion of judge Cooly about the "right to independence". Cases involved the famous Prince Albert, in the opinion of Warren and Brandeis, on one hand, the traditional protection of intellectual property rights are connected with a part of personal privacy interests. And outside except the property interests protected by copyright, the right of personality also is due to

[1] Prince Albert v. Strange (1848) 41 Eng. Rep. 1171 (Ch.).

protection.

In 1903, after the publication of the "right to privacy", the New York state legislature allowed to appeal on the grounds of privacy violation because personal name, image or image is used for commercial purposes[1].

But regardless of in the article of "the right to privacy", or prosser's summary of the type of infringement related to personal information, ignored the existence of another theoretical foundation of the privacy interests of the United States, namely, the principle of confidentiality. The principle of confidentiality, in the legislation and judicial practice, there are four kinds of concrete forms, first, is the system of privilege, such as marital privilege system, the doctor's privilege, the prerogative of the priest, when the court of a personal investigation and evidence collection, husband and wife, between the doctor and the priest does not bear making adverse testimony obligations. On the other hand, in the judicial precedents of the United States in the 19th century, based on position and trade ban, information acquired in lectures and partnership relationship, because of the violation of the principle of confidentiality, and the parties have caused the injury behavior of the proposed tort, it can also obtain the recognizance of the court. Third, since Britain inherited in order to prevent the extortion of the aristocracy, Britain has developed extrotion to prevent those noble servant and lover and other related people to threaten to sell or open the scandal while trying to get sealing fee. This has gained the USA law conviction. The fourth case, government has the confidentiality obligation to citizens because of the behavior of public power to obtain personal information of citizens. No matter what kind of situation, the principle of confidentiality exists for a long time in the protection of personal affairs in the USA and plays an important role. And these things haven't been reflected in the summary of the tort behavior by Prosser. So were neglected in theo-

[1] Zhang Mina, *The Comparative Study of the Right of Privacy—France, Germany, the United States and Other Countries of the Right to Privacy*, Zhongshan University Press: China, 2013, p. 61.

ry. And the principle of confidentiality based on the identity relation is exactly the important way that we regulate the protection of genetic information, this relationship because of identity is an important starting point that we build the right subject and right system.

Hence, it is precisely these neglected situation in the legal system that has very important significance to the protection of genetic information. This means that, for the protection of individual genetic information, except based inviolable dignity views on individual freedom, the protection on the individual emotional interest, and another important reason is the specific interpersonal relationship obligations. Although in the case of 19th century and 20th century, these specific interpersonal relationship was mainly summarized as trust relationship, this information, to a large extent, is contained and regulated by theory of intellectual property, but not the type of genetic information protection of the initial choice of tort law about the violation of personal affairs but this trust relationship opened the length and breadth of vision that we study the genetic information for the protection of the rights. Be specific to the protection of genetic information, the protection we have talked about in the past, if is only limited in "the right to the world", then against research institutions and their staff, commercial organizations and public power external stakeholders, the right to obtain the information is confrontational. And ignore the fact that to specific family, because of the existence of common genetic information, there exist conflict between privacy interests. Except to make measure between the individual interests and rights of life and health, there is also a case that it does not threat to life and health, or the threat is not urgent, if family members learned its genetic information or disclosed their information how to regulate the conflict, we need to set up from the standing point of interpersonal relationship. And that's more than just a consideration of human dignity.

In America, in the protection of the judicial practice of information, an important principle is "reasonable expectation of privacy". The principle of justice was founded in the bill of coexisting opinion in the case of Harlan in Katz v. United States[1]. The criterion of reasonable expectation of privacy is: "individual citizens have objective expectation of privacy to be searched, and the public think the expectation of privacy is rational, even if the searched place does not belong to the stipulation of the fourth amendment of the American federal constitution[2], then the behavior also constitutes a search behavior."[3] In the case of Katz, police installed monitoring equipment outside the telephone booth, although the telephone booth is not a traditional privacy protection object, but due to the anticipation of public recognition of the parties in use of a telephone booth, then its interests should also be maintained. However, when the criterion of "reasonable expectation" is quoted by another famous case, the focus of our concern is whether the information detected by emerging technologies is also contrary to the citizen's personal information maintenance for the expectations. In the Kyllo v. United States case[4], the focus of the parties to dispute is whether the use of new equipment to detect kyllo can become a reasonable basis for the search. In 1991, two American police officers Elliltt and Dan Haas suspected Danny Kyllo at home to grow and the indoor cultivation of marijuana must use bright light exposure, because the house will certainly has an anomaly heat of the daily life. So two policemen carrying a thermal imager and scanning the front and back sides of the house for few minutes, and thermal imager analysis showed that, Danny kyllo houses really had more heat than surrounding buildings. According to the evidence of "informant reports, property bills and a thermal imager analysis results", federal court signed searching a warrant for Danny kyllo. The re-

---

[1] Katz v. United States, 389 U. S. 347, 361, 19 L. Ed. 2d 576, 88 S. Ct. 507.

[2] Fourth Amendment to the United States Constitution: The right of the people to be secure in their persons, houses, papers, and effects, against unreasonable searches and seizures, shall not be violated, and no Warrants shall issue, but upon probable cause, supported by Oath or affirmation, and particularly describing the place to be searched, and the persons or things to be seized.

[3] Zhang Mina, *The Comparative Study of the Right of Privacy—France, Germany, the United States and Other Countries of the Right to Privacy*, Zhongshan University Press: China, 2013, p. 61.

[4] Kyllo v. United States, 553 U. S. 27 (2001).

sults showed that he really used light to grow marijuana at home, thus he was prosecuted.

Danny filed for the exclusion of illegal evidence, that the police violated his privacy. But whether the evidence is illegal became the most important focus in the case during the process of the trial, the appeal and the federal court back to in the retrial. There were also great controversy between the judge. Because housing is certainly most expected private place of citizens, and the traditional areas protected in the constitution of the United States of America (the modified four case).

But the police did not break into Kyllo's home, but only to carry out inspections on public roads. The judge in the judicial trail also stressed that even if home privacy are protected but you can't request law enforcement police blindfolded when passed private residence. And now, if you don't use a thermal imager, but only according to observation of the eyes, whether the police were able to make sure that kyllo was growing marijuana at home. So the problem became a search problem. In another judicial case, Dow case established standard that if the police use technology to enhance the sensory effects, it can constitute a search. Because although the law never asked the police to be blindfolded when passing the private residence, but if there is no thermal imager, people's senses may not be aware of illumination caused by abnormal heat; if there is no infrared observations, eyes also can not pierce the tile barrier. And in fact, the reason why people have great demand on housing is to isolate the area of its own which is independent from public life. Therefore, these new science and technology is opposite to the meaning of housing, so may infringe on the interests of the citizens. While the case is experiencing a destiny of appealing, remanding for retrial, it is also based on the maintenance of public interest, protection of private rights, and the result of gaming the predict factors considerations of emerging technology. The reasonable expectation of privacy, indeed becomes an important standard of protection of individual citizens in the face of the current technology development.

Although, in the administration of justice, application of technology has been widely recognized by the public, advances in technology, brought about the result that legislation or public awareness, always lag behind the novelty of technology at the time of the incident. Genetic engineering and interpretation of genetic information is the most significant achievements of this era. When we think that gene detection threshold cost is the obstacles of gene equality that the understanding of the genetic information is just a game for the rich. Gene technology company has substantial reduction in the cost of genetic testing with the individual gene sequencing $1000 mark below, meaning not only the popularization of science and technology, also the cost of violation of privacy is greatly reduced. The progress of gene information technology is also quietly brought people's traditional sense, even the effect that traditional science and technology can not achieve. At this time, there may exist conflict between subjective interests and objective privacy. One of the tasks of law is to protect individual rights to safeguard the interests of privacy and leave a predetermined reasonable expectation standard for the application technology.

2. 2. Trust Benefits of Genetic Information

Individual genetic characteristics, except for a distinctive external performance, such as hair color, skin color, height and weight and so on, and the mystery of the containment in the interior of the body, only after genetic testing and scientific analysis, can be identified as legal genetic information. The legal protection of trust interest means that when we analysis report that in scientific research institutions and specific personnel for detection and analysis, it does not mean the loss of confidentiality. Trust interest also means that, even if the individual genetic information in the database also involves the interests of the individual.

In Britain the protection of the interests of the individual information in the common law can also be traced back to Prince Albert v. strange case, but in the UK, the interpretation of the case exist difference from the points of Warren and Brandeis. The British court mainly from the the perspective of illegal trust responsibility principle to apply in this case. In the case, they focused on the special relationship between the plaintiff and the defendant. Apart from Prince Albert v. Strange, the case also has important significance for the protection of privacy. First of all, in 1913 Ashburton V. Pape case[1]. The case established the principle that "people are prohibited public the secret information through illegal means or leaks in the process of transmission." In 1948 Saltman Engineering Co. V. Campbell Engineering Co. case[2] is formally established the position of breaching the trust as a common law appeal, while the case has gained the recognization that the parties in violation of the trust responsibility not only exist in the contractual relationship. Even if as long as the third party recognize their duty to carry out the duty of confidentiality in obtaining this information, then behavior of making public may constitute an infringement. In 1969 Coco. v. A. N. Clark[3] established three basic conditions to constitute a violation of trust responsibility for infringement; first, the open information must have confidentiality; second, the defendant has the duty of confidentiality; third, the unauthorized disclosure would cause actual damage to the parties. What need mention is about the three basic conditions, the confidentiality of the information is not exactly equal to the privacy, which means that the personal information which has told the specific individuals do not necessarily lose the interests of the information and whether the information belongs to a public area is not a exact standard. And British law stipulated that even the information are obtained from the public domain, but if the collation and processing behavior also damage the right of privacy, still can constitute infringement of the elements as violation of trust responsibility.

In English law, a landmark event was the pass of 1998 British "bill of rights"[4]. The bill of rights requires English law to protect the human rights which

---

[1] (1913) 2 Ch. D. 469 (C. A.) (U. K).
[2] (1963) 3 All. E. R. 413 (1948) (U. K.).
[3] Coco. v. A. N. Clark (Eng' rs) Ltd, (1969) R. P. C. 41 (U. K.).
[4] Human Rights Act, 1998, c. 42.

is established by European Convention on human rights to expand the scope of protection of the rights of their own. And the article eight of European Convention on human rights is exactly about the right of privacy [1]. Although prior to this, the British was always avoiding to use the general concept of the right of privacy, the process of European integration driven by the European Convention on human rights requires that English law must protect privacy interests through a reasonable way. With the "bill of rights" promulgating and being effective, appeals for infringement of "personal information interests" are gradually increased. This requires that in the judicial practice, these situations shall be protected. Although the British Parliament and jurists, intentionally or unintentionally avoided that the independent status of the right of privacy law did not get recognized by legislation or judicial practice. With frequent cases, English law by expanding breach of trust principle can contain genetic information, including personal information protection. A typical case is about the famous movie star Douglas and Zeta Jones v. Hello case [2]. Although this is news media against the will of the star couple, a typical violation of the privacy because public their wedding scene, the court still prefers to explain that all guests to participate in the wedding should clear know that the couple don't want to be published in the media. But the protection for the rights of the information gene takes the path of British law which is different from the United States The difference is mainly expressed in the following aspects: first, the starting point for Warren and Brandeis to occupy the mainstream of the United States information protection is "the right of independence", also called a standing point of individualism. What they are trying to protect is the inviolability of the individual dignity and personal feelings; and the starting point of the protection of English law of genetic information is social public relations, focusing on the relationship of trust between people in the protection. The

[1] European Convention on Human Rights, Art8 (1), Nov. 4, 1950.

[2] Douglas v. Hello! Ltd, 92001) Q. B. 967 (2000) (U. K).

theoretical origin is the principle of confidentiality, the protection path is to expand the protection of genetic information including the breach of trust. Second, the property of information is essential in the protection of genetic information for the United States, namely the information once it has been leaked will cause shame of the party, causing the parties of unpleasantness and discomfort. But in British the violation of trust relationship between infringement is not valued attributes of the information itself, but the obligation of confidentiality that the parties bear. As long as the information is confidential and should be compromised, it may cause infringement. This principle makes English tort law in the face of such as commercial institutions, for the legitimate purpose and means of access to personal information of citizens, but may leak out, will be able to provide more effective protection. Such as for the main body without pathogenic and defective genes, the analysis of their genetic information is not able to cause the parties to be shame and discomfort, but still belong to the category that should be protected. Third, in the United States, if the disclosure of genetic information is only within a handful of people and this does not constitute infringement of the rights, but in the UK, which is also violating the obligation of confidentiality. On the other hand, the two theories about opening a distinct locations is that the attitude of English law on the protection of genetic information, gives us a great inspiration, that if our information is revealed to a particular person, it does not mean we give up our right of genetic information. And this is precisely the point of view that can reveal the genetic information we are confronted with the problems in the protection of privacy. First, because of the condition of medical, insurance or education and other reasons, our genetic information is submitted to a specific agency, but it does not mean that our information can be used for the second time without our consent. Especially it should not easily become the basis against our interests. At the same time, in the field of scientific research, acquisition of this special information does not mean we give up the related personal interests and property interests. Of course, it does not mean that individuals lost the further qualification of sharing interests. And in family relationships, family members share the similar or partly same genetic information because of kinship, whether can also be regarded as a kind of special relationship of trust. Then the behavior of opening their own genetic information whether violates the confidentiality obligations of other family members.

Thus, it became a new perspective for us to examine gene privacy protection taken the right of privacy as a tool.

2. 3. The Property Interests of Genetic Information

Through the analysis of the phenomenon of right related to the attribute of genetic information. The analysis involved in genetic information rights is not only the personality interests. With a detailed analysis of the the oretical system of personal information protection, it is not difficult to find that except starting from the angle of individual autonomy, the reasonable expectation interests and trust interests related to genetic information is the important content of the protection, and has a profound the oretical origin, but one of the key issues of the protection of genetic information which can not be avoided is the maintenance of the property interests. An important point of view is the care of genetic information in those implied in the advanced theory and practice of legislation. Except for the tradition of protection of personal information in the United States and the United Kingdom, Wagner, an American scholar, in his "French privacy theory development"[1], James Q. Whitman "Two kinds of privacy culture in Western country: dignity and freedom;[2] and Jeanne M. Hauch "the French privacy protection——the existence and development of the theory of Warren and Brandeis in France"[3]. Three articles reveal the content which also can not be ignored. According to Professor Whitman's view, an important tradition

[1] Wenceslas J. Wagner, "The Development of the Theory of the Right to Privacy in France", 1971 *Wash. U. L. Q.* 45, pp. 45 -69.

[2] James Q. Whitman, "The Two Western Cultures of Privacy: Dignity Versus Liberty", 113 *Yale L. J.* 1151 (2003 -2004), pp. 1171 -1172.

[3] Jeanne M. Hauch, "Protecting Private Facts in France: The Warren and Brandeis Tort is Alive And Flourishing in Paris", (1993 -1994) 68 *Tul. L. Rev.* 1219, pp. 1219 -1301.

in France is that nobles often asked people to open their property status, including real estate security, and this unopened appeal often got approved legally. The maintenance of personal information, is parallel with the French Revolution and the French news media. With France's introduction to the freedom of the press system and freedom of speech are advertised and development, the lives of citizens by the threat of danger is also getting increasing high. Jacobin believed that it was essential to consolidate the achievements of the French Revolution, protecting civil rights, not only to maintain the freedom of speech, but also to safeguard citizen's private life from infringement. In his initiative, article 17 of French 1791 constitution provides that "behavior of defamation or infringe upon the private life of others, based on the prosecution of others, the actor's behavior should be punished."

At the same time the French constitution of 1791 information only changed for the protection of the aristocrats, subject has extended to the scope of ordinary people, consolidating the protection of the rights of citizens by legal reform.

Another French scholar who made outstanding contributions to the protection of privacy rights is Pierre Paul Royer Collard. In 1819, he puts forward a famous judgment that their private lives must be separated by a fence. This thesis not only become the French direct reference to the the oretical basis in the later judicial practice of 19th century, but also directly promote the development of mainland right of privacy theory.

Thus, the maintenance of citizens always existed in the social life of France, and form finalized as an important outcome of the French Revolution in the constitution of 1791. But for a long time, on the study the process of history in France privacy protection has not been clear, the main reason is that, as an important content of civil law, the origin of the right of privacy is largely dependent on academic works of civil law scholars. The French civil law scholars did not show great interest on the right of privacy theory. But this does not mean that the personal information and the protection of private life in the French civil law is a blank concept. And it is precisely that, for a long time, based on the French Civil Code provisions can be included on all citizens information and involving the interests of protection, which makes the concept of the French Warren and Brandeis era unlike in America which seemed so urgent. Because in France, according to article 1382 and article 1383 of civil code of 1804, the party shall bear the liability because of the behavior of fault tort . The regulation of the provisions is very broad and is not limited to specific forms of infringement. As long as the actor has subjective fault, the objective existence of the damage, both damage to the material also includes the spirit damage and the fault and damage shall bear tort liability. Therefore, the infringement of citizens' interest which are attached to the genetic information, of course, should also bear responsibility. So the French civil code successfully include infringement case of genetic information. In judicial practice, article 1382 and article 1383 of civil code of 1804 also became the legal basis for information protection. Two representative cases are the Rachel Affaire case[1] and Alexandre Dumas Prere case. In the case of the Rachel Affaire, Rachel is a star, when she lay dying, the family asked the photographer to take pictures of her in order to make mark, but the photos soon were exposed to an painter, painter took the photos as blueprint drawing and sold to the public. Rachel's family appealed because the violations of private life, and the claim to collect and destruct the photos of the painters was upheld. And in the case of Alexandre Dumas prere, the indecent photographs captured by the plaintiff and his star girlfriend was sold by photographer without consent, the claims of infringement also was upheld by judges. What is worth noting in two cases is that although all claims are caused by the photo, but the reason why plaintiff was upheld is the violation of personal private life rather than a claim based on the right of portrait. Therefore, the judicial practice of the French proved that citizens can safeguard their interests of data information based on the fault tort liability provisions of civil code of 1804.

---

[1] T. P. L. de la Seinem June 16, 1858, D. P. III 1858, 62.

On the other hand, the interests of data information can not only be maintained by the civil code general tort clause. Because the upsurge demand of the French privacy interests is inseparable with expansion of news reports, French legislators enacted 1868 news law, news reports will constitute a crime if the public private life of others. Also in the field of judicial practice, to the private life, especially the crime involving information not only shall bear the criminal responsibility of press law, and the victim was also able to claim civil damages. Although the French nobility have the right for the maintenance of its reputation, in a duel or legal way to prevent personal information has been made public or threatened by others. And in 1804 civil code of France, it also can be included through the fault liability, protection of the interests of the information . But contemporary maintenance of personal information is different from the way of 1804 French civil code. Before the 1950s, demands were required to satisfy the elements of general fault tort liability, which was the most important, in the lawsuit, the plaintiff need to prove that the defendant has subjective fault. There is no doubt that the plaintiff in the maintenance of information in the course bear the burden of proof. Therefore, with the development of economy, in the progress of the news publishing industry, frequent cases of privacy, in judicial practice, courts began to give up to making the plaintiff to prove the other party was faulty.

A decisive case became a time point to divide the traditional and contemporary protection. In 1954 Dietrich affair case, the plaintiff contacted with publisher for his reminiscences, but in the process of negotiations, publishers published the plaintiff's private life in the journals in form of papers publication. Plaintiff told the court. There were two meanings of the case: first, the court did not require the plaintiff to prove the fault of the defendant and the defendant hereby shall assume the tort liability. Second, judgment in the case not only maintained the mental damages suffered by the plaintiff, because of the adverse effects caused by the leakage of private things and family details without the permission of the plaintiff and public the memoirs earlier that expected, which made plaintiff lose benefits of his memoirs so the decision was that the defendant bear 120 million francs for the compensation for property damage. Thus, it opened a new pattern of contemporary French privacy protection. In the justice of France, it gradually no longer requires the plaintiff to prove the defendant's fault, and in some violation, not only for property compensation, the most important is to eliminate the damage and potential damages. So the court began to seizure or destruct the items of the defendant that may cause damages to the plaintiff, such as newspapers, magazines and videos, while such an approach has no enactment of basis. Therefore, there are some academic questions.

The article 9 of the French civil code of 1970 appeared. The 1st provision of article 9 provides that any person enjoys the respect of the rights of private life. The scholars nowadays, except for the provision of the EU data protection of the maintenance of genetic information, the Civil Code Article 9 is an important basis for the gene information maintained. Article 9 (2) states: when the defendant who has infringed upon others' right to privacy, the judge has the right to order the perpetrator to assume the liability for damages, in addition to the judge has the right to take various measures to avoid or end behavior of violations, such as seizure relating materials, confiscation of infringing someone's material or other measures; as in the case of emergency, the judge of the court can take these measures. So the maintenance of personal genetic information is not limited to the mental compensation caused by the leakage of information, the most important is to take measures to the property.

What need explanation is whether it is based on the civil code of 1804 1382 and 1383 the protection of personal information or according to the 1970 Civil Code Article 9 give the including genetic information, privacy protection, behavior person shall assume the tort liability. But this kind of tort liability is obviously different. First, in the judicial practice, the plaintiff shall bear different burden of proof. Nowadays, the personal information infringement on the plaintiff only need to prove the damage to their own interests. But in the protection of the traditional mode of existence, damage, fault, causality between behavior and damage are essential in lawsuit proof; second, the protection of personal information in the civil code of 1804 is included in the general tort liability, so the defendant shall bear the liability, and in ninth section of the contemporary civil code, this responsibility is the presumption of fault; third, the measure for legal compensations is extended from the past simple damages to other coercive measures and damages for infringement and remedies the law of personal information. Thus, the French model based on legislation and judicial practice was established, this protection mode of traditional information to protect the interests have both inheritance and breakthroughs, undoubtedly increasing the strength of the protection of personal information right, and leaving space for the protection of genetic information property interests.

While in the field of traditional continental law, including some civil law scholars in France believed that Article 9 the French Civil Code was a kind of protection of personality rights. But for the French genetic information and the protection of the rights of property, we should inspect the content of judicial practice.

In France, there are not only the trail because privacy interests are infringed and loss of publication access to material compensation, but the trail that after the death of the French President Francois Mitterrand, the courts have barred the doctor in a short period of time from announcing president medical information for a publication. Therefore, the maintaince of genetic information, medical information, personal information, can not simply be limited to the narrow private place, also can not be blind whether is only limited to personal interests. The protection of genetic information in France is to maintain the value of freedom. This freedom, can be in private space and in public places, this freedom is the principle of equality, without any distinction between politicians, celebrities and the general public. This freedom can produce the delay of requirements of personality protection, with the need to share interests.

Therefore, France has provided a good path for gene information protection. As a kind of personal information, genetic information reveals an individual's health status, predicting future risk of disease, can be the most secret personal data, related to the way treated by society. Of course, the dignity of the individual is related. At the same time, along with the increasingly close combination of genetic information and medical technology, genetic information as one of the most important medical information, it can be incorporated into the scope of protection of civil law. And this protection mode, on one hand, it can cause great importance for genetic information in daily life, on the other hand, it may trapped into the flaws of gene determinism. It is not because gene can interpret human nature, but genetic information is related to human dignity, which is as same as other medical records. It should be responsible for citizens to decide whether public or not. The property of interest related to genetic information has got the establishment of relief means and the subject of genetic information rights can share the interests according to individual freedom.

3. the Composite Property of Genetic Information Right

In biology, genetic information decides the individual identity and it can be inherited between generations. On ethics, human genetic information are concerned with human dignity, and moral values. In the field of social development, with the promotion of progress of medical research of gene technology, business interests groups have got huge profits, promoted and accomplished the special value of genetic information in the social cognition level. However, with the many legal interest disputes, apart from fully understanding the objective existence of the special nature of genetic information, the key problem is to answer how to recognize the legal statue of genetic information, or how to define the genetic information right in the legal system.

3.1. The Traditional Defining Manners of Rights Property in Private Law

There are three main methods to define the attribute of genetic information in the traditional private law, as personality, property and intellectual property. The advantage to define the genetic information as a kind of personality is able to make the individual right existing on the genetic information included by the traditional personality rights protection system, which has a positive meaning for eliminating the genetic discrimination and safeguarding freedom, equality and human dignity. However, protecting the genetic information in the personality right system will face the constantly trouble that the commercialization and patenting of genetic information has become a fact. Personality right can not solve all the problems posed by the use of genetic information. As the Moore case, the deny of property attributes of genetic information may eventually lead to the negative maintenance of the personality. To protect the genetic information as the property legal interests or the subject of intellectual property rights is in line with the needs of attribution of participants in gene technology. However genetic information, hosted the function of "life code", "medical records" and "family and community mark" and so on, after all, is not equivalent to a general "objects". Therefore, the transfer, distribution and resource allocation of genetic information is facing with the ethical dilemma different from the general materials. The approach of overlay the "property legal interests" on the "personality legal interests" can only result in avoidance of the nature of the problem.

As it is said in the book "The control of pest arm and tat liver" by Yan Jue'an, in terms of gene technology, "Conceptual tools, analytical framework and theoretical supposition recourse by the traditional jurisprudence have been fundamental challenged." [1] Research on genetic information, must break through the traditional private research category. While the core issue of legal regulation of genetic information is not to answer the ethical proposition "to protect or to restrict", but to provide more operational guidance norms. Accordingly legal regulation of gene information must firstly return to the level of origin of the right to be explored, breaking the strict boundaries of public and private research binary mode in order to meet the challenges

---

〔1〕 Yan Jue'an, *The Control of Pest Arm and Tat Liver: The Exploration of Legal Theory and Bioethics*, Peking University Press, 2006, p. 110.

of the right content, the right jurisdiction and the forms of right competence by the genetic information.

In existing Chinese studies on the legal status of genetic information, Moore case has been used as a classic case that vividly reflects the information asymmetry between researchers and sample donors to lead unfavorable protection of individual right. The focus of right protection is concentrated on how to define the property of the portion separated with Moore's body. In response, pundits made selection as "body" or "property" of protection path.[1]

3. 2. The compound maintenance of multiple interests

The event which reflects more vivid particular value of genetic information and has more influence on biology practice is about the use and controversy of "Hela Cells". The famous Hela cells derived from an African – American woman Henrietta Lacks' cervical biopsy blocks. Unfortunately, Henrietta Lacks couldn't escape and died of the cancer disease. But Hela Cells nurtured by her biopsies could be the first human cells to grow on their own in the laboratory. And it no doubt could be called be the cell which made the largest contribution to the human medical technology. For example, Scientists developed the polio vaccine and found the telomerase based on the Hela Cells, almost at the same time more than 400 detect fragments of Hela cells existed in laboratories all over the world, thousands of research papers were both in English and Chinese literatures……

Although Henrietta has been dead for a long time, What let Lacks family concern was that in 1971, the journal "Obstetrics and Gynecology" published the article illustrates the relationship of Hela cells and Henrietta Lacks and the naming scheme was widely used by the authority magazine such as "Nature" and 'Science". Thereafter, the Lacks family had a doubt about the use of Hela Cells and family members

[1] There are two typical opinions, first is Professor Wang Zejian tended to choose regard the separation part of Moore's body as a part of body and use a personality right, second, Professor Chen Wenyin choose to protect the separation as a property.

were collectively collected blood samples. However, with the situation of Henrietta herself, the family members couldn't get introductions, also out of informed consent. These days, Lars Steinmetz team in European Molecular Biology Laboratory in Heidelberg, Germany published an article entitled "transcriptome profiles of the genome of Hela Cell line" and "Nature" magazine prepared to publish more detailed research on "genome map Hela cells"[1], which let Lacks family believe that it made their loved Henrietta's medical records public, and the privacy of family members thus has been compromised. Although the Lacks family envisioned a variety of solutions with regard to economic compensation, patent protection and access to justice program to safeguard its interests, finally, they made a choice of agreeing the sequence results stored in the nonpublic databases which was built on the strict access restrictions as the consequence to satisfactorily resolve the distribution for the Hela Cells. The satisfactory consequence would be another contribution by the Lacks Family. And the event would undoubtedly open the new perspectives of legal research. Research focus turned from the distribution of defining the genetic information as property or body to how to control the circulation of such a kind of information that affected the individual identity, family relationship and medical, technological and even public interests, among different subjects and in different fields.

Genetic information not only has the objective particularity, but also has effect on the traditional protection system of private law. Therefore, protection of the rights of individual genetic information must be consistent with this objective particularity, and maintain multiple interests of the individual's body, information, liberty and property. By the investigation of analysis of right attribution and historical exploration of the right connotation, establishing genetic information rights protection mode is able to meet their diverse right subject and balance their multiple benefits in a differ-

---

[1] See "Nature: Hela Cells Dispute Has been Satisfactorily Resolved", contained Web Biological Exploration, http://www.biodiscover.com/news/research/105563.html (last visiting date 2015.03.01).

ent field, and lay the foundation for the construction of reasonable legislative protection mode.

The essence of construction of right is to resolve conflicts of interest among different subjects in different jurisdictions. The subject of genetic information right, based on the difference of genetic information particularity, could be divided into two parts. The first part is based on the personal identification and family and community correlation, so its related resource allocation must consider the main interests of the individual, family members and community. The construction of the second part of subject is based on the function of genetic information to predict disease risk, as well as technical reproducibility, thus the flow of genetic information often involves the different users of biological samples and information, such as doctors, researchers, commercial investors and officers. The collection, distribution and use of genetic information along with the different needs of its different subjects, occurs in the family, medical, research, employment, education, insurance, management of government and other public areas. With making a generalization of different needs of genetic information in a high degree, we find that it includes four elements, obtaining genetic information, the circulation space of genetic information, the self – control of genetic information and information assets. It can be clear that the construction of genetic information right is a multi – dimensional work. Breaking through binary boundaries of "person" and "Matter" in the traditional private law has become a necessity. It should be noted that the right we are talking about, as the theory of Hohfeld, is tending to be a bundle of right[1], which must abide the different value such as the freedom, equality, property and so on and make a balance in different values.

For the protection of genetic information, British and American scholars prefer pragmatic stance, and they pay attention to the specific regulation in using the genet-

[1] Wesley Newcomb Hohfeld, "Some Fundamental Legal Conceptions as Applied in Judicial Reasoning", *Yale Law Journal*, Vol. 23 (1913).

ic information in the concrete places. Some scholars have proposed a "right to privacy" or "equal rights" as an analytical tool, but emphasize at the same time the study of these rights must respect specific allocation rules in different areas. In civil law system, the use of related concepts of genetic information rights is relatively vague and scholars tend to analysis the legislation of right rather than defining the right connotation. The contents of "European Convention on Human Rights and Biomedicine" and "The Chapter of Fundamental Rights" explicitly prohibit genetic discrimination, and "Data Protection Directive" has empowered the right of controlling personal information from the standpoint of the individual human right. Detailed anglicizing of information right provided by the European Union, the right of "informed" or "knowledge" on individual genetic information, is a personality right, which can't include the property interests.[1] Facing with the emerging new situation and new problems and the impacted traditional protection mode, scholars who only recognized the personality interests on the genetic information have proposed a close relationship between personality and property and the property right of genetic information also help safeguarding the human dignity. In domestic literatures, whether using the "the fundamental human rights of gene," "genetic privacy", "genetic property", "genetic intellectual property", " gene right as the right of personality" or by "genetic rights" which includes all the concepts of analysis mode[2], it can be seen in the system and outstanding literature studies, involved interests of genetic information is not a single, should fully protect their information, body, liberty, and

〔1〕 See Karl Larenz, *The General Theory of German Civil Law*, translator Wang Xiaohua, Law Press: China, 2002, pp. 159 – 173.

〔2〕 To investigate the gene right as a basic right, see Zhang Xiaoluo, "Study on Gene Right", Wuhan University doctoral thesis in 2010. To use the genetic privacy, see Luo Shenghua, *Legal Protection on Gene Privacy*, Technology Press: China, 2010. To investigate the gene property, see Han Ying, "Establishment of Property Rights of Human Genes", *Hebei Law Science*, Vol 12 (2008). To investigate the gene right as a kind of personality right, see Wang Kang, Private Regulation on Gene Right, Fudan University doctoral thesis in 2012. To study the specific content of gene right in different level, see Qiu Geping, *Study on Human Gene Right*, Law Press: China, 2009.

property values.

If defining the genetic information as a kind of specific information in the legal norms, the protection requirements of obtaining, circulation and controlling the information can be met by the protection mode of establishing a specific right of genetic information which reflects the specific and hierarchical tendency of information right. The core issue is the right property to cover the multiple benefits of genetic information. Therefore, genetic information right is not only experiencing the universal concept of consciousness, it is also a theory to get the process of creation and system construction.[1] We can not deny that, the essential origin of genetic information right is consistent with the personality right in protecting the self - determination, which is focused on the individual autonomy. However, with the arrival of the information age, science and technology development not only break the traditional mode, but also are gradually weakening the emphasis of individual autonomy, and emphasizing the flow of genetic information. The development and use of genetic information put forward new demands on theory of rights. As the study of Professor He Jianzhi, the key problem of disputes about genetic discrimination in privacy is the right of information[2]. The autonomy of body is not the specific problem of genetic information, while genetic information operation in practice, on the other word, processing of information, precisely is distinct from human organ and tissue samples' problems. The focus transferred from the determination to processing of information, is the value of compound interests as the right attributes of genetic information right. Apart from the effect on the individual interest by privacy, the right protection takes account of the effect by right on the individual in community and social value of information.

Looking at the regulation and protection of the rights of genetic information, we

---

[1] See Xuliang, "On Privacy Right", Wuhan University doctoral thesis in 2005.

[2] See He Jianzhi, *Study on Genetic Discrimination and Legal Measures*, Beijing University Press: China, 2006, p. 55.

find that the main line behind it, namely the compound of interest. The law must consider the needs of different interests of different stakeholders, respect the specific guidelines of allocation of resources in different areas of society, recognize the benefits of coexistence with the game and make a rational choice in order to realize the social cooperation and individual development.

LEGAL CONNOTATION AND BOUNDARY OF GENETIC AUTONOMY RIGHT: IN THE BACKGROUND OF POPULATION GENETIC DATABASE

WANG Kang[1]

Abstract:

For a natural person, he has personality right of self - determination gene, namely, genetic autonomy right. By its very nature, the basic legal value of gene is freedom, and the natural legal connotation informed consent. With influences of human dignity, freedom of others and no injury principle, genetic autonomy right is by no means to be exercised absolutely freely. Anyone who reveals his genetic information will affect others. In a wider circle, population genetic database complicates informed consent. The core value of genetic database is social justice rather than unilateral altruism. Constructing trust relationship based on communication in the situation of risk will solve the conflict between individual autonomy and public interests better and promote the proper use of population genetic database as public - interests resources. In genetic era, we should reconstruct legal idea and legal order of social coexisted responsibility.

Keywords: genetic autonomy right; population genetic database; genetic information; self - determination; social coexisted responsibility

Raising Problems

In May 14, 2013, Angelina Jolie published an article My Medical Choice in the New York Times. It goes that Angelina's mother died from cancer at 56 and Angelina was diagnosed as pathogenic gene BRCA1 carrier, which led the probability of mam-

[1] Professor, Shanghai University of Political Science and Law, PhD in Law.

mary cancer and oophoroma to 87% and 50% respectively. For the sake of health, children and family, she chose to undergo a double mastectomy to lower the risk. Angelina said, "As soon as there's a problem that I have to face, I decide to deal with it positively and to reduce suffering risk to the lowest". Besides, she also proposed that women with history of disease in the family should take preventive measures to their best. [1] Angelina's medical choice and her bravery to reveal her suffering were applauded by the public. What's more, Angelina was chosen as cover story of New York Times, and the press named it as "Angelina Effect". "Angelina Effect", however, was more than praise, and there were some doubts as well. Besides the doubt of enormous benefits behind Jolie [2], revelation of her own genetic information, resulting in exposing genetic privacy of other family members, was also met with incredulity. Is it necessary that other family members who share the same genetic privacy with Angelina agreed before self - determination? Should we set some legal limits for genetic privilege?

This special right could be called genetic autonomy, and we should observe it in a wider circle. For the purpose of understanding the association between gene and disease and other external factors like lifestyle or environment, or realizing social management or other public interests, some research projects, "population genetic

---

[1] Angelina Jolie, "My Medical Choice", *New York Times*, May 14, 2013.

[2] Mike Adams, EXPOSED: Angelina Jolie part of a clever corporate scheme to protect billions in BRCA gene patents, influence Supreme Court decision. http://www.naturalnews.com/040365_Angelina_Jolie_gene_patents_Supreme_Court_decision.html.

database" or "biobank" [1] are promoted in numerous countries and regions. As for the construction of population genetic databases in different countries and regions, state - level projects in Iceland, the UK and Japan are more striking, and genetic database of mainland and Taiwan are in operation. [2] Population genetic database in this paper is a genetic data consisting of human tissue sample (e. g., blood), genetic material, analytical data and other related information including lifestyle, medical records and history of disease in the family, which is to understand the association and acting mechanism of disease, gene, environment and lifestyle by means of population - based studies. Population genetic database includes not only unitary genomic data but also other datas, Iceland, Estonia, the UK, Japan and China are

---

[1] Besides these two titles, it can be called "human genetic base", "group database", "human genetic base", "human body bio - database, etc. OECD considers it as HGRD, see OECD, Creation and Governance of Human Genetic Research Databases, OECD Publishing, 2006. We choose the notion of "population genetic database" in this paper, then Estonian Genome Project, UK Biobank, CARTaGENE, UmanGenomics, Genome Database of Latvia Population, Genome Institute of Singapore, Autogen Limited, Biobank Japan Project, etc. See Jocelyn Kaiser, Population Databases Boom, From Iceland to the U. S., Vol. 298, No. 5596 (2002), pp. 1158 - 1161; Margit Sutrop, Human Genetic Databases: Ethical, Legal and Social Issues, Trames: Journal of the Humanities and Social Sciences, Vol. 8, No. 1/2 (2004), pp. 5 - 14. The comparison see Anne Cambon - Thomsen, "The Social and Ethical Issues of Post - genomic Human Biobanks", *Nature Reviews Genetics*, Vol. 5 (2004), pp. 866 - 873; Brian Salter & Mavis Jones, "Biobanks and Bioethics: The Politics of Legitimation", *Journal of European Public Policy*, Vol. 12, Issue 4 (2005), pp. 710 - 732; Lin Ruizhu, "The Development of Database Protection System and Its Application to Mass Genetic Database", *Tsinghua Law and Policy of Science and Technology*, Vol. 2, No. 3 (2005), pp. 153 - 188; Li Daixuan, Analysis of Legal Issues of the Establishment and Management of Human Gene Database, Master' Thesis at College of Law at NTU, 2009.

[2] Typically in the Mainland is 2004 Kadoorie Study of Chronic Disease in China, lasting 10 to 15 years. On Taiwan Biobank, see Fan Jiande, Liao Jiacheng, "On the Legal and Ethical Planning of the Establishment of the Mass Genetic Database in Taiwan", *Taipei University Law Review*, Vol. 68 (2008), pp. 95 - 149; He Jianzhi, The Ethical, "Legal and Social Studies of the Biological Database in Taiwan: Review and Policy Analysis", *Taiwan Law and Policy of Science and Technology*, Vol. 5, No. 1 (2008); Liao Peishan et al, Public Attitude towards Gene Database in Taiwan, *Taiwan Law and Policy of Science and Technology*, Vol. 5, No. 1 (2008), pp. 167 - 193; Wu Tingfeng et al, Public Communication about Human Biobank: A Preliminary Study of Experience in Taiwan, published in the Second Annual Conference of Taiwan Society of Technical and Social Studies: "Differences and Connections: STS, Engineering and Society", *Kaohsiung* (Taiwan), 2010, 5, 15 - 16; Tang Shumei et al, A Survey of the Public Attitude towards Gene Database in Taiwan on Ethical, Legal and Social Issues, published in the Second Annual Conference of Taiwan Society of Technical and Social Studies: "Differences and Connections: STS, Engineering and Society", *Kaohsiung* (Taiwan), 2010, 5, 15 - 16.

implementing or have implemented the project. [1] Different from general medical research projects, population genetic database is of forward goal – design plan and large – scale sample sources, which cover lager – scale population and also gear the need to the future (even more generations). Thereforc, subsequent ethic, legal and social problems are unique and complicate informed consent. When collecting population tissue samples and related information, research direction is more or less set and it's impossible to foresee all researching and applying goals. Therefore, some inevitable problems emerge——what should the participant be informed before consent? Need some express or written consents or special ones? The consent is from an individual or the group? What if some agree while some are opposite? Should majority – decision democratic mechanism be applied? How to define group? Finally, the core problem: how to use genetic autonomy right?

Legal Connotation of Genetic Autonomy Right and Limits

a. Legal connotation of genetic autonomy right

Genetic autonomy right is a kind of personality right that a natural person has for self – determination on gene.

The core of genetic autonomy right lies in self – determination. Self – determination, an ethic principle, means that autonomous action is not in control of others, [2] divided in weak one and strong one, in weak sense, self – determination equals to rejection to external intervention; in strong sense, however, it is regarded as a component of internal value. In law, early, Justice Cardozo, J. stated the notion of "right of self – determination" in 1914. [3] Right of self – determination (right of

---

[1] Other genetic databases, such as Human Genomic Database or DNA Sequence (American Genbank, European EMBL, Japanese DDBJ, etc.), criminal DNA database and other databases of animals and plants are not included in this paper.

[2] Tom L. Beauchamp & James F. Childress, *Principles of Biomedical Ethics*, 6th Edition, New York: Oxford University Press, 2008, p. 126.

[3] See Schloendorff v. Society of New York Hospital, 211 NY 125, 105 N. E. 92, 133 N. Y. S. 1143 (1914).

self - determination of private things) is a new personality right deriving from the principle of autonomy in private law, i. e. , a rational man can determine his own thing by himself. It is himself that determines his future and fate to achieve the development of personality. [1] According to American academics, self - determination equals to independence generally, and in the sense of familiar philosophy, it means the ability to realize what the rational man should do and freedom to do out of unlawful intervention. Moreover, self - determination spreads out in four senses as follows: ①capability to manage oneself; ②actual conditions and merits of self - management; ③perfection of characteristics (in a state of ideal); ④sovereignty of managing oneself (in the range of moral of someone). The core of self - determination is the right to choose and determine how to use his property, which individual information to reveal to others and so on. In short, the most basic right of self - determination is the right to determine the lifestyle. [2] The positive effect of genetic autonomy right, a microcosmic sign of the right of self - determination in genetic era, is that individuals could make decisions freely in the field of private affairs on their own genetic personality interests.

An individual with genetic autonomy right is able to control and use his specific gene appropriately in two dimensions——personality and property. Essentially, gene of human beings is such a kind of personality interests, also considered as personality property, that the subject possesses "quasi - ownership" of genetic material and information. And then he owns some legal effects including undoubted power to control and deal, privilege to use and share benefits, request for excluding intervention and immunity to withstand the request from interests - related person etal. First of all, in the aspects of collecting, saving, maintaining and ruining of genetic material, the subject could make decisions after informed completely. Therefore, positive effect of

[1] Liu Shiguo, Study on Issues of Newborn Personal Rights, *Legal Forum*, Vol. 6 (2011).

[2] Anita L. Allen, Richard C. Turkington, *Privacy Law: Cases and Materials*, Chinese Democratic and Legal System Press, 2004, pp. 366 - 367.

genetic autonomy right could be used to deal with tort, conducted and conducting, including hunt pirate and fraud of gene. International Declaration on Human Genetic Data Article 2 emphasizes that anyone given informed should freely make express agreement of an individual to his or her genetic data being collected, processed, used and stored. Chapter 9 of Data Processing, Data File and Individual Liberties in France [1] is about processing personal data for the purpose of medical research. Article 56② provides, "where the research requires the collection of identifying biological specimen, the informed and express consent of data subjects must be obtained prior to the implementation of data processing". Then in the aspect of testing and revealing genetic information, subjects could make self – determination of whether to test and reveal and the way, range and degree and ask for excluding improper intervention or approaching from the third person (e. g. , family member, employer, insurer) by using genetic autonomy right. Even, Swiss Federal Constitution provides that a person's genetic material may only be analyzed, registered or disclosed. [2] Lastly, in the sense of genetic controlling of life potential (i. e. , using personality of gene) via genetic medical technology, it is the genetic autonomy right that subjects use to control genetic condition of unborn children properly in condition of human dignity and inter – generation justice. For example, breed a baby free from carrying gene of some serious genetic disease via genetic diagnostic technology before embryo implanting. This rational right of controlling life potential is different from reproductive right. For the latter, its effect only influences first – generation babies, while the former may cover all the offspring.

Genetic autonomy right means free development of genetic personality, making self – determination of control and rational use of genetic material or information, self –

[1] Data Processing, Data File and Individual Liberties (Amended by the Act of 6 August, 2004 relating to the protection of individuals with regard to the processing of personal data).

[2] "A person's genetic material may only be analyzed, registered or disclosed with the consent of that person, or if a statute so provides." Article 119 (2) f, The Swiss Federal Constitution.

control of life potential in condition of human dignity and inter - generation justice and denying genetic "hunt", "pirate" and "fraud". By its very nature, the basic legal value of gene is freedom, and the natural legal connotation informed consent.

b. Legal limits of genetic autonomy right

However, genetic autonomy right is by no means to be exercised absolutely freely, and self - determination, either. Were there no limits of genetic autonomy right, it would be a catastrophe for the society characterized by compromise, co - decision and common liability.

Above all, the first boundary of genetic autonomy right is human dignity. Unexpectedly, though Yan Juean puts forward the notion of genetic right, he suggests everyone possesses genetic right of information other than right of controlling life potential of reproductive cells. And the potential personality should be entitled to best possible conditions of life, which withstands and exceeds right of controlling life potential. [1] This opinion is based on the field of constitutional rights, which is certainly meaningful. If genetic autonomy right includes right of complete reproductive autonomy, modifying reproductive gene would have legitimacy. However, embryo or fertilized egg is regarded as "patient" actually, and it may violate inter - generation justice in ethic and law because of probable invasion of fundamental human rights and dignity of the posterity. Universal Declaration on Bioethics and Human Rights Article 11 provides for this. [2] Proper control of life potential could be confined to avoiding serious disease only. No "design" is permitted for nonmedical purpose like selecting intelligence, gender and appearance. Convention on Human Rights and Biomedicine confines the intervention seeking to modify the human genome to be undertaken only for preventive, diagnostic or therapeutic purposes and only if its aim is not to intro-

---

[1] Yan Jue'an, *Half Alive: On the Legal Normalize to Life, Regulation of the Liver of Rats and the Arm of Insects: Essays on Jurisprudence and Bioethics*, Angle Publishing Co. , Ltd, 2004, p. 38.

[2] Article 11: No individual or group should be discriminated against or stigmatized on any grounds, in violation of human dignity, human rights and fundamental freedoms.

duce any modification in the genome of any descendants. Also, it provides the use of techniques of medically assisted procreation shall not be allowed for the purpose of choosing a future child's sex, except where serious hereditary sex – related disease is to be avoided. [1] If we could choose or modify some genes that decidc characters of children, the posterity would become naked products and tools. Even if we don't need to think about the damage of genetic diversity now, we can imagine that all the human being will lose nature as a man.

Then, the second boundary of genetic autonomy right is others' freedom. Kant considers coexistence of will and others' freedom as general principle of right. [2] In genetic medical era, use of self – determination is more influenced by diverse complicated factors derived from social system. Therefore, some self – determination should be made only by interests – related group for common interests or respective intention. In special condition, we must respect and protect others' right to choose. Universal Declaration on Bioethics and Human Rights Article5 [3] and

[1] Art. 13 & 14, Council of Europe, Convention for the Protection of Human Rights and Dignity of the Human Being with Regard to the Application of Biology and Medicine: Convention on Human Rights and Biomedicine (European Treaty Series – No. 164, Oviedo, 4. IV. 1997).

[2] Kant, *The Philosophy of Law: An Exposition of the Fundamental Principles of Jurisprudence as the Science of Right*, Commercial Press, 1991, p. 41.

[3] It provides that "the autonomy of persons to make decisions, while taking responsibility for those decisions and respecting the autonomy of others, is to be respected. For persons who are not capable of exercising autonomy, special measures are to be taken to protect their rights and interests".

Article6 [1] provide it in detail. According to contemporary Confucian bioethics, Fan Ruiping stresses familism in self – determination in China (compared with western individualism) and family decision are prior to individual determination. [2] This interpretation reflects harmony and tolerance that self – determination needs before the freedom of others.

Finally, the third boundary of genetic autonomy right is "the principle of nonmaleficence". "The principle of nonmaleficence", a crucial principle of self – determination, could be traced back to On Liberty (J. S. Mill). Mill holds that the only intention to exercise right properly to others is to prevent from damage. Living in society, everyone should observe the bottom line when referring to others without any damage to others' interests and try to protect the society or its other members from injury. [3] Therefore, exercising rights must be harmless to others. Universal Declaration on Bioethics and Human Rights Article4 is so provided. [4] Even unimagina-

---

[1] It provides that "1. Any preventive, diagnostic and therapeutic medical intervention is only to be carried out with the prior, free and informed consent of the person concerned, based on adequate information. The consent should, where appropriate, be express and may be withdrawn by the person concerned at any time and for any reason without disadvantage or prejudice. 2. Scientific research should only be carried out with the prior, free, express and informed consent of the person concerned. The information should be adequate, provided in a comprehensible form and should include modalities for withdrawal of consent. Consent may be withdrawn by the person concerned at any time and for any reason without any disadvantage or prejudice. Exceptions to this principle should be made only in accordance with ethical and legal standards adopted by States, consistent with the principles and provisions set out in this Declaration, in particular in Article 27, and international human rights law. 3. In appropriate cases of research carried out on a group of persons or a community, additional agreement of the legal representatives of the group or community concerned may be sought. In no case should a collective community agreement or the consent of a community leader or other authority substitute for an individual's informed consent".

[2] Fan Ruiping, *Contemporary Confucian Bioethics*, Peking University Press, 2011, pp. 29 – 39. Advantages of familism see pp. 72 – 73.

[3] See Anita L. Allen, Richard C. Turkington, *Privacy Law: Cases and Materials*, Chinese Democratic and Legal System Press, 2004, p. 364.

[4] It provides "In applying and advancing scientific knowledge, medical practice and associated technologies, direct and indirect benefits to patients, research participants and other affected individuals should be maximized and any possible harm to such individuals should be minimized".

ble, a deaf lesbian couple in America wants to give birth to a deaf baby. [1] The law and moral rule about right of self-determination must be set in orbit pursuing welfare of all the human beings.

Situational Presentation of Genetic Autonomy Right: Take Population Genetic Database as an Example

Genetic autonomy right contains self-determination that emphasizes on informed consent. "Informed consent" was first used in a case from California State Court of Appeal in 1957 [2], and the decision clarified that only after receiving information from doctors could consent of patient enter into force. Plenty of clinical trials against human rights were disclosed in 1966 [3], which resulted in a focus on informed consent. WMA Declaration of Helsinki proclaimed the informed consent principle expressly when it was revised in 1975. It is generally assumed that informed consent consists of four basic elements——full notification of information, understanding of information, ability to consent and free expression of consent. In genetic medical era, these elements are still the core of genetic autonomy right, in the new situation, taking population genetic database as an example, with new problem and special sense.

a. Subject of self-determination

On subjects of consent [4], whether consent of certain genetic group (though confining the range of this community is a problem, the smallest unit is family undoubtedly) is more necessary than private law and it's better to consider it as an important issue of constitution. Most international norms, such as Nuremberg Code and

---

[1] M. Spriggs, "Lesbian Couple Create a Child Who is Deaf Like Them", *Journal of Medical Ethics*, Vol. 28, No. 5 (2002), pp. 283-283.

[2] Salogo v. Leland Standalone Jr. University Board of Trustees, 317 P. 2d 170 (Cal. App. 1 Dist. 1957). Case and analysis see Duan Kuang, He Xiangyu, "Physicians' Duty of Informing and Patients' Commitment", Liang Huixing ed., *Civil and Commercial Law Review*, Vol. 12, Law Press, 1999, p. 152.

[3] Henry Beecher, *Ethics and Clinical Research*, *New England Journal of Medicine*, Vol. 274, No. 24 (1966), pp. 1354 - 1360.

[4] I don't discuss the individuals without the ability to consent there in this paper.

Declaration of Helsinki, all promote informed consent, and Universal Declaration on Bioethics and Human Rights Article 13 encourages interactions between individuals. In HUGO Statement on Proper Conduct of Genetic Research, informed consent could be made at the level of individual, family or community. When setting population genetic database, consent of individual, family or group is an inevitable issue. Iceland and the UK adopt the way of consent of individual, while Tonga consent of family when collecting tissue samples. [1] In China, Interim Instrument on Management of Human Genetic Resources Article 12 provides permission process about international cooperation project of human genetic resources, and sets credential of consent from supplier of human genetic resources material and his relatives as a requirement, i. e. , China adopts consent of family indirectly. As a genetic community, a magnified family, genetic group, must consider the whole interests positively. In the sense of special representation rights, there are two requests of the genetic community—— one is to protect the group from negative effects of internal dissent; the other is to prevent from influence of external decisions. Will Kymlicka name these two internal restrictions and external protections? [2] If there were representation mechanism, "civil right of differential groups" would play a role in deciding whether to participate in population genetic database, especially coordinating internal different ideas and enhancing capability of negotiation to protect genetic right of group. Because there is no group presentation mechanism, when it comes to reality, however, individual consent is generally enough in China, and only in special conditions, family consent, for instance, of common interests is necessary. Involving crucial personality interests of an individual, in case of "the majority autocracy", majority - decision

---

[1] Ellen Wright Clayton, Informed Consent and Biobanks, *Journal of Law Medicine and Ethics*, Vol. 33, Issue 1 (2005), pp. 15 - 21.

[2] The idea of special representation rights is produced in the political process of western democratic states, i. e. , offer certain amounts of seats in the legislature to the weaker, which is a correctly political response to oppression or systematical shortages. See Will Kymlicka, *Multicultural Citizenship: A Liberal Theory of Minority Rights*, Shanghai Century Publishing Group, 2009, pp. 40 - 45.

democratic mechanism should be applied prudently when determining whether to participate in population genetic database, even if in the genetic community. In other words, both consent from community and individual are needed, the former covering common genetic private and the latter referring to individual freedom.

b. Forms of self - determination—general consent or re - consent

On contents, due to a long - term research, time - oriented biomedical requirement leads to incomplete informing to the participants before consent (including use, interests and risk). [1] In the light of the reality, informed content is often insufficient, which is opposite to the requirement that doctors should give complete and sufficient information to the supplier before receiving consent. [2] However, genetic medical research and trial are different from routine medical treatment and the latter is of lower risk and has fewer questions. So, doctors have to take more responsibilities to inform the supplier. In the research of population genetic database, informed content could only be predictable information, in complete there of Ariticle 4 of Regulations on the Keeping and Utilization of Biological Samples in Biobanks in Iceland lists informed items: the nature of the biological sample; security measures applying to the taking and preservation of the biological sample, and the nature of personal identification pertaining to them; to whom the biological sample will be entrusted; the donor is free to grant authority for the preservation of the biological sample in biobank and refusal to grant such authority will have no effect upon the donor's legal rights; the donor can withdraw consent for gathering a biological sample, or cease participa-

[1] Levitt & Weldon, "A Well Placed Trust? Public Perceptions of the Governance of DNA Databases", *Critical Public Health*, Vol. 15, No, 4 (2005), pp. 311 - 321; Fan Jiande, Shi Zhiyuan, "On the Impact of the Development of Ethnic Studies on Legal Norms in the Biomedical Field: Cross - strait Biological Database as A Instance, Ni Zhengmao, Liu Changqiu eds", *Law of Life: Essays on the International Symposium on the Development of the Science and Technology of Life and the Construction of Legal System in* 2007, Heilongjiang People's Publishing House, 2008, pp. 170 - 192.

[2] Ellen Wright Clayton, "Informed Consent and Biobanks", *Journal of Law Medicine and Ethics*, Vol. 33, Issue 1 (2005), pp. 15 - 21.

tion in a scientific study at any time; *rules of procedure of the biobank.* [1] Although it provides like this, information is still incomplete. However, it doesn't mean that "consent" on the basis of this is improper. The only problem is how to define the range of "consent" that has been given, namely, whether the situation is included that we cannot predict the purpose and use of research definitely at the time that consent is given.

If the answer is yes, the consent will be a kind of "general consent", i. e., the consent contains supplying biological sample for uncertain research in the future. [2] If otherwise, "re - consent" will be necessary, namely, before new research, the participant must give a consent again. These two forms reflect different ideas and the answer is not chosen by ticking simply. It depends on whether the individual data is identifiable, if so, "re - consent" is necessary; otherwise, it is "general consent". Considering the general goal of public interests of population genetic database and the nature of prospective study, we should show preference for "general consent" in condition of following the rules of informed consent and anonymity. Ethic Regulation 2000 in Japan suggests that whether re - consent is necessary depends on the duration of the sample and the condition of the supplier, and the regulation is supported by Ethic Review Committee. IMRC Principles of Bioethics of Human Body also provides that reusing the sample, which is anonymous, should be approved by Ethic Review Committee. The reasons for advocating re - consent are as follows: it reflects the respect for autonomy right of the sample supplier; mitigating potential injury to supplier; reducing stress; preventing researcher from overusing of genetic sample and data. However, re - consent is facing realistic problems. Firstly, the researcher himself could not predict how to use genetic sample in the future, so, it's unrealistic

---

[1] Liu Hong'en, "Genetic Technical Ethics and Law: the Self - discipline", *Heteronomy and National Norms of Biomedical Studies*, Wu - NanBook Inc., 2009, pp. 12 - 13.

[2] Alice Hsieh, "A Nation's Genetic Database", *Columbia Journal Law & Social Problems*, Vol. 37 (2004), pp. 359 - 411.

for the supplier to make a choice on the uncertain research, meanwhile, it brings lots of troubles to the supplier. Secondly, getting in touch with sample supplier will increase extra financial burden. Thirdly, if the supplier doesn't consent to the reuse of sample, the validity of research data will be in danger, the efficiency reduced thereby. In conclusion, "general consent" is necessary. In an investigation to Chinese Human Genome Research Center (Beijing & Shanghai) and more than 10 researchers from Human Genetic Institute and committees from Ethic Review Committee, 300 survey participants are divided in the attitudes tore - consent: 59.7% of them are in favor of it, 32.7% contrary and the rest have no idea. [1] As the report by one commission of American Academy of Sciences[2] states, to demand the research participant to consent to uncertain use of those identifiable DNA samples in the future is unacceptable no matter in ethic or law. In 1999, a report published by NBAC[3] suggested that in particular conditions, we could relax rigid requirement for informed consent to some extent after approval from Ethic Review Committee. Fewer potential risks as anonymous form results in to the supplier, re - consent can be omitted with the approval of Ethic Review Committee. However, in most conditions, anonymous sample has little research value. After approval and review of Ethic Review Committee, requirements of the immunity informed - consent contains as follows: anonymous sample is not relevant to identifiable information; the supplier can withdraw the sample at any time; immunity of re - consent is a better choice for the supplier. [4]

Biobank Law in Iceland provides that collecting tissue samples must observe the rule of complete informed consent, rather presumed consent in former Sanitation Di-

---

[1] Zhang Xinqing, "Human Genetic Database in China: Ethical Issues and the Attitude of Scientists", *Science*, Vol. 2, 2007.

[2] Committee on Human Genome Diversity, Commission on Life Sciences, National Research Council, Evaluating Human Genetic Diversity, Washington, DC: National Academy Press, 1997, p. 65.

[3] National Bioethics Advisory Commission, Ethical Issues in Human Stem Cell Research (September 1999).

[4] See Zhang Xinqing, Fan Chunliang, "Discussion on Ethics, Management and Policies and Regulations on the Construction of Human Genetic Database in China", *Scientific Culture Review*, Vol. 3, 2007.

vision Database Law. Article 7 of Biobank Law fixes that before the conservation of human tissue sample in biobank, "full informed consent in the state of freedom" is indispensable. The supplier must be completely informed of the purpose, use and potential risks and benefits of sample collection and the long - term conservation of the sample in biobank to be used according to this law. Only after being fully informed of these items, will the consent in written form conform to statutory requirement, "full informed consent in the state of freedom". Furthermore, because sample supplier has the right to withdraw the consent whenever necessary, the biobank has to ruin the tissue sample after the withdrawal of the consent, except for substances and data that have been produced from the study. But even so, identifiable data must be deleted thoroughly and reuse for further research is forbidden. As provided in Regulations on the Keeping and Utilization of Biological Samples in Biobanks, using tissue sample for research is a kind of genetic research rather than physiological or cellular study. Informed consent from the sample supplier is necessary, especially in condition of tracing back to identity, which could not make presumed consent as the substitute. [1] This provision is of great significance for domestic regulations. Anyway, informed consent is a fundamental ethic and principle of law of genetic medical research and we must assure complete informed voluntariness of the sample donor.

c. The core of self - determination—the known and unknown of research outcome

Another important question—whether to offer the supplier a feedback particularly informing test or research outcome? [2] As talked above, knowing risk of future disease is not always good, on the contrary, we can get benefits living without know-

---

[1] Liu Hong'en, *Genetic Technical Ethics and Law: the Self - discipline, Heteronomy and National Norms of Biomedical Studies*, Wu - NanBook Inc., 2009, pp. 12 - 13.

[2] Early arguments see K. A. Quaid, et al., "Disclosure of Genetic Information Obtained Through Research", *Genetic Testing*, Vol. 8 (2004), pp. 347 - 355; G. Renegar, et al., "Returning Genetic Research Results to Individuals: Points - to - consider", *Bioethics*, Vol. 20, No. 1 (2006), pp. 24 - 34.

ing existence of the risks. Therefore, whether the feedback is needed should consider the autonomy right of the participant. If the participant exercises the "known" right, researchers should make an express request initiatively and inform the participant of the right to get an feedback, especially about possible risks expressly, in an document on informed consent at least. As for the "unknown" right, it is presumed by implied form.

As provided in regulations of the UK, offering a feedback is not an obligation of population genetic database and the researcher. The Iceland suggests that the participant does not have right to obtain any information about research outcome, except that he could request the report of treatment measures and safe standard of individual data. The provision about not offering research achievement is also in Estonia Human Genes Research Act[1], which contains that the participant can apply for use of individual data (not including relatives' data) conserved in database. Latvia is inclined to offering feedback to the participant. [2] Some instruments of international organizations have preference for supporting the feedback. Convention on Human Rights and Biomedicine, passed by Council of Europe in 1997, states that everyone is entitled to know any information collected about his or her health, however, the wishes of individuals not to be so informed shall be observed. In exceptional cases, restrictions may be placed by law on the exercise of the rights contained in paragraph 2 in the interests of the patient. [3] HUGO holds that participants should also receive information about the general outcome (s) of research in understandable language. The ethical advisability of provision of information to individuals about their

---

[1] Estonia Human Genes Research Act (2000).

[2] Margit Sutrop, "Human Genetic Databases: Ethical, Legal and Social Issues", *Trames: Journal of the Humanities and Social Sciences*, Vol. 8, No. 1/2 (2004), pp. 5 – 14.

[3] Art. 10, Council of Europe, Convention for the Protection of Human Rights and Dignity of the Human Being with Regard to the Application of Biology and Medicine: Convention on Human Rights and Biomedicine (European Treaty Series – No. 164, Oviedo, 4. IV. 1997).

results should be determined separately for each specific project. [1] International Declaration on Human Genetic Data (UNESCO, 2003) also provides when human genetic data are collected for the purpose of medical research, epidemiological research population – based genetic studies or genetic testing, the information provided at the time of consent should indicate that the person concerned has the right to decide whether or not to be informed of the results. [2] Some academics state expressly that it is against the principle of respecting liberty of medical research that the UK biobank doesn't offer any feedback information. An investigation conveys that most participants prefer to accept the feedback on physical health. [3] In fact, genetic autonomy right can explain psychological risk due to informing appropriately. But, for population genetic database, though the participant might be of genetic information on the risk of serious diseases, the information is not of specific clinical value, and offering feedback individually will increase process and timing and financial cost. [4] Generally speaking, on the basis of the nature of medical study, unless participants are discovered of some serious disease and medical therapy exists, [5] it's hard for population genetic database to predict the risk and injury of disease of the participant and to construct such full "communication room" [6] as concrete relationship between doctors and patients, impossible to perform legal obligation of informing there-

[1] HUGO Ethics Committee, Statement on Benefit – sharing (2000).

[2] UNESCO, International Declaration on Human Genetic Data (2003).

[3] Carolyn Johnston & Jane Kaye, "Does the UK Biobank Have a Legal Obligation to Feedback Individual Findings to Participants?", *Medical Law Review*, Vol. 12 (2004), pp. 239 – 244.

[4] Some states, these costs are not excuses for refusal to inform the participants of crucial information. See Carolyn Johnston & Jane Kaye, "Does the UK Biobank Have a Legal Obligation to Feedback Individual Findings to Participants?", *Medical Law Review*, Vol. 12 (2004), pp. 239 – 244.

[5] In fact, MRC points out that researcher should design a plan to inform the participants and consider the beneficial information. And the participants are entitled to choose whether to get information, except that research results indicate serious disease of the participant and it can be curable. See Medical Research Council, Guidelines on Human Tissue and Biological Samples for Use in Research (2001).

[6] Liu Yuanxiang, "The Informing Duty of the Research Achievements of the National Human Genetic Database", *Law and Life Science*, Vol. 4, 2008, pp. 31 – 44.

of. Moreover, in term of public interests, it is not fair that population genetic database, not pertaining to hospital and not about individuals, undertakes this responsibility. Once the participant asks for this, offering feedback will be an obligation. However, considering principle of caution, researcher needs to remind the participant of the "unknown" right before offering feedback. When covering interests of third person, more complex, more ethic and legal considerations are needed to coordinate possible conflict of rights.

Coordination of Genetic autonomy right and public interests

As other genetic medical research and technology application, population genetic database is full of various uncertain potential risks in setting and operating. For participants, population potential risks of genetic database include physical, psychological and information risk: physically risk, an acceptable and foreseeable risk, is direct injury of the body (e. g. , improper collection); psychological risk, preventive beforehand, pertains to emotional injury, such as discomfort resulting from that disclosed information (genuine or wrong) is not expected or ethic and legal factors. Information risk, hardly predictable, refers to direct injury in using genetic data (e. g. , improper revelation of genetic information) or indirect discrimination to the individual and group. [1] In specific condition, the individual can refuse to participate in the research, but, public - interests goal of population genetic database will not come true and the conflict of genetic autonomy right and public interests, individual consent as the core, is inevitable.

Traditional bioethics is based on individual, while population genetic database is about a large - scale group. Different notions and rules suit different situations. How to coordinate the conflict of private right (e. g. , right of privacy, secret and the

---

[1] See Niu Huizhi, The Study and Discussion on the Qualitative Analysis of Potential Risk Issues in Human Genetic Database and Risks Management Models, Qiu Wencong ed, Biennial Journal in 2007 of the Development of Science and Technology and Legal Norms: The Legal Construction of the Risks of Public Health, Preparatory Office of the Institute of legal science of Academia Sinica, 2008, pp. 143 - 196.

known and unknown) and public interests and unity (e. g. , the obligation of participation, informed – notification) is the most important challenge. There are no general ethic and legal rules to support universal application of population genetic database, although the UNESCO, HUGO, WHO and WMA have focused on the challenges. [1] In Statement on Human Genomic Databases, HUGO Ethics Committee stresses that the choices and privacy of individuals, families and communities with respect to the donation, storage and uses of samples and the information derived therefrom should be respected; informed consent may include notification of uses (actual or future), or opting out, or, in some cases, blanket consent; mechanisms should be established to ensure respect for such choices. Essentially, the best benefits of population genetic database lie in medical progress rather than commercial benefits and core value is social justice not unilateral altruism. To reach the goal, the balance of individual freedom and public interests is necessary, and the principle of trust might be the bridge.

Then, confronted with two modern contradictions inserted in genetic medical situations deeply—trust VS risk and opportunity VS danger [2], how to constitute the trust? Because of the basis of social participation and open expression on coping with risks, resulting in public trust, sufficient understanding and communicating will play an important role in the operation of population genetic database. In Statement on Human Genomic Databases, HUGO Ethics Committee stresses that individuals, families and communities should be protected from discrimination and stigmatization; prior consideration should be given to the possible negative socio – economic effects, if any, of the collection, sharing, and publishing of the data. Therefore, the respect of individual autonomy contains that participants are able to understand a series of risk talked above completely and that the participants are titled to choose whether to

[1] Margit Sutrop, "Human Genetic Databases: Ethical, Legal and Social Issues", *Trames: Journal of the Humanities and Social Sciences*, Vol. 8, No. 1/2 (2004), pp. 5 – 14.

[2] Anthony Giddens, *The Consequences of Modernity*, Yilin Press, 2000, p. 130.

get study results on their own genetic information. On coping with population genetic database, some suggests to build a "partnership" mechanism——by means of continually valid communication and participation in executing and supervising under the mechanism of population genetic database, we should convert the donors into positive participants as "partners" from being passive in long-term relationship, and build "substantial trust relationship" between donors and researchers to decrease potential risk derived from interests' interweaving of population genetic database and to reach the goal of mutual benefits. [1] "Partnership" mechanism is actually a concrete practice, but how to implement it remains to be discussed.

Conclusion

In genetic era, analysis on genetic code brings about new questions one after the other, which lead to new inspection of traditional legal ideas without any doubt. With influences of human dignity, freedom of others and no injury principle, genetic autonomy right, informed consent as the core, is by no means to be exercised absolutely freely. For instance, Angelina, a specific individual, discloses her own genetic information often refers to interests of others (especially family members). Then, a proper negotiation and conversation are needed. In a wider circle, population genetic database complicates informed consent. Constructing trust relationship based on communication in the situation of risk will solve the conflict between individual autonomy and public interests better and promote the proper use of population genetic database as public-interests resources. On the basis of this, we should reconstruct legal idea and legal order of social coexisted responsibility.

[1] Niu Huizhi, The Study and Discussion on the Qualitative Analysis of Potential Risk Issues in Human Genetic Database and Risks Management Models, Qiu Wencong ed, Biennial Journal in 2007 of the Development of Science and Technology and Legal Norms: The Legal Construction of the Risks of Public Health, Preparatory Office of the Institute of Legal Science of Academia Sinica, 2008, pp. 143-196.

Part Two　Biotechnology and Bioethics

PATENT AND BIOTECHNOLOGY

Alexandra Mendoza - Caminade[1]

Abstract:

The drug's legal status has changed considerably towards its free appropriation by the way of the patent. This commodification of the pharmaceutical product is now widely accepted by most states and by international texts, particularly in the World Trade Organization, which have chosen this orientation. Various justifications have been advanced to justify the interest of this merchantability of the drug. But this industrial and commercial logic of pharmaceutical companies is difficult to conciliate with the interests of public health and human health, particularly for patents on biotechnology.

It is in this study to see if the current legal model promotes a balance between private and public interests. The conclusion is an imbalance in favor of the industrial logic and the patent. Facing certain excesses, rebalancing mechanisms must be put in place and such as, the Supreme Court of the United States participates in The Myriad case of this movement by limiting the scope of patentability in biotechnology.

Keywords: drug; pharmaceutical; biotechnology; patent; public interest

Today, the tension between the patent and the drug is so numerous that the legitimacy of pharmaceutical patent is contested at both national and European level and internationally.

Long, the drug was designed as a non - market value before becoming the sub-

---

[1] Associate Professor of Law, Toulouse 1 Capitole University, College of law - business law center, Director of Master Intellectual property.

ject of an industry. Indeed, the health product[1] has become a commodity almost as another[2] that trades beyond the state borders[3]. If the reservation through the patent has become the rule, admission of the patent on pharmaceuticals is recent[4]: the refusal to patent pharmaceuticals continued until an order of 4 february 1959[5], before a law of 2 January, 1968 only includes the drug in the common law of the patent[6]. Legislative developments were completed with a law of 13 July, 1978[7] and the article L. 611 – 16 of the Code of Intellectual Property embodies the principle that patentability[8].

At the international level, the protection of the pharmaceutical product by the

---

[1] The term drug must be understood in a broad sense refers to drugs which are, strictly speaking, pharmaceuticals and products derived from living matter. According to the article L. 5111 – 1 of the Code of Public Health, drug "means any substance or drug composition presented for treating or preventing against human and animal diseases, as well as any substance or composition can be used in humans or animals or that can be administered to them, to establish a medical diagnosis or to restoring, correcting or modifying physiological functions by exerting a pharmacological, immunological or metabolic action".

[2] Special regulations exist concerning in particular the marketing and circulation of such a product. About the specifics of the drug traffic and the National and Community administrative regulation which is the subject: F. Chaltiel, *Free Movement of Drugs: Recent Developments*, PA October 27, 2005, n° 214, p. 11.

[3] A drug may be of worldwide interest when populations are affected by a pandemic. Thus, a biopharmaceutical company that is developing a drug to fight against this pandemic has a potentially global market: about this global dimension of medicines: C. Jordan – Fortier and I. Monk – Dupuis, *Globalization, Law Competition and Protection of Health: The Case of Pharmaceuticals, in Globalization and Competition Law*, W. Abdelgawad (under the direction of), Litec 2008, p. 439.

[4] Tracing the legislative developments in this area: F. de Visscher, *General Interest and Patents: Some Thoughts about Legislative Developments, In The Public Interest and Access to Information on Intellectual Property*, Bruylant 2008, p. 93, spec. p. 98.

[5] The order of February 4, 1959 and a decree of 30 May, 1960 set up a special patent of the drug.

[6] In European law, it is the Munich Convention of 5 October, 1973 revised the November 29, 2000 which clarified that possibility: see new section 53 c of the Convention. In France, it is the law n° 2007 – 1475 of 17 October, 2007 (OJ October 18, 2007), which authorized the ratification of this act, and Decree n° 2008 – 428 of 2 May, 2008 (OJ May 4, 2008) publication covers the Act revising the Convention on the Grant of European Patents.

[7] J. Azéma, Is there still a specificity of pharmaceutical patent?, JCP ent. 1990, II, 15744.

[8] According to the article L. 611 – 16 of the Code of Intellectual Property, "are not patentable methods of surgical or therapeutic treatment of the human or animal body and diagnostic methods practiced on the human or animal body. This provision does not apply to products, in particular substances or compositions, for the implementation of these methods".

patent is allowed and the patent is widely used. In the framework of the World Trade Organization, the agreement on aspects of intellectual property rights relating to trade, said TRIPS, of 15 April, 1994 requires member states to protect the patented inventions in all fields, including pharmaceutical[1]. Thus, the model of the patent and intellectual property is becoming universal[2]. In addition, the field of pharmaceutical patent has continued to expand and include new health products[3]. Thus there's an inexorable progression of merchantability health products and pharmaceutical patents.

If the patent has become essential for the marketing of drugs is due to the incentive for research[4] through the reward offered to the creators with the monopoly on the creation for twenty years, allowing the recovery of costs research and development of innovative companies[5]. The patent is traditionally presented as a contract between the Company and the inventor[6] whereby the Company grants an operating monopoly to the inventor on innovation to encourage research and to encourage researchers to disclose their innovations[7].

---

[1] Article 27, 1 of the agreement. There are significant commercial pressures developed against countries that do not yet have a system of effective protection of intellectual property countries. On this point: M. Barré - Pepin, *Globalization of the Patent System and Counterfeit Drugs in the Medicine and Person*, I. Dupuis (under the direction of), Litec 2007, p. 185, spec. p. 191.

[2] In this sense: J. Foyer and M. Vivant, *Patent law*, PUF, 1991, p. 6.

[3] Referring to the expansion of intellectual properties and the phenomenon of appropriation of living: JM. Bruguiere, *Intellectual Property Law*, Ellipses, 2005, p. 4.

[4] Citing "an effective mechanism for economic incentives": J. Schmidt - Szalewski and JL Pierre, Industrial Property Law, Litec 4th ed. 2007, n° 35; J. Azéma and J - C. Galloux, *Industrial Property Law*, Dalloz, 6th ed. 2006, n° 157.

[5] On the assessment of the average cost of drug development: E. Combe and E. Pfister, *Patent and Drug Prices in Developing Countries*, Propr. ind. July 2003, p. 272.

[6] Since the 19th century, the term of the contract has been in use to describe this special relationship linking the inventor to the company: see in particular E. White, *Treaty Infringement*, Paris, 1855, p. 463 and 464; more recently, see C. Nozaradan, Patent and General Interest, in Patent, Innovation and General Interest, B. Remiche (under the direction of), Larcier 2007, p. 445, spec. p. 454.

[7] On the subject of the legal action: JM. Mousseron, Treaty Patent, Litec 1984, n° 11. The World Trade Organization also states that "the way intellectual property is protected may also serve social objectives": WTO : TRIPS and pharmaceutical patents, www. wto. org, September 2006.

The link between innovation and patent is so important today that creating and marketing a drug cannot be considered without patent. However, industrial and commercial logic of the pharmaceutical companies must be reconciled with the interests of public health and human health[1], particularly in terms of patents on biotechnology. The law of patents is intended to strike a balance between private and public interests[2], as recalled in the domestic and international laws[3]. But the balance is hard to find: the patent appears too powerful to the detriment of the public interest, and that hegemony is highly controversial (I), so that the stop blow by the Supreme Court of the United States to this expansion of biotechnology patents appears relevant (II).

I Excesses of biotechnology patents

A - General criticisms

Among the classic criticisms of the patent, it would be a disincentive for research and diversity of this research[4], and impede excessive competition. In addition, the privileged position of the patented lead to set an unusually high drug prices[5].

---

[1] In 2008, the European Commission has launched an inquiry into the pharmaceutical sector in the European Union to determine if actions in competition law should be instituted. This survey and other initiatives of the European Commission are "to provide European patients with safe, effective and affordable medicines, while creating a business environment that stimulates research, boosts a useful innovation and support the competitiveness of the industry". Among the steps in this investigation, there has been a preliminary report of 28 November, 2008 which opened a consultation period culminating in a final report of July 8, 2009: European Commission: http://ec.europa.eu/comm/competition/sectors/pharmaceuticals/inquiry/index.html.

[2] About finding balance in terms of patents, see G. Van Overwalle, *The public interest, the public domain, the commons and the law of patents, in The general interest and access to information on Intellectual Property*, Bruylant 2008, p. 149. See also: Consumer, drugs and pharmaceuticals, G. Michaux (under the direction of), *European Journal of Consumer Law* n° 2 -3/ 2009, p. 229.

[3] Article 7 of the TRIPS Agreement states that "the protection and enforcement of intellectual property rights should contribute to fostering technological innovation and enhance the transfer and dissemination of technology, to the mutual benefit of those who produce and those who use technical knowledge and in a manner conducive to social and economic welfare, and to a balance of rights and obligations".

[4] In this sense, B. Bergmans, *Protection of Biological Innovations—a Comparative Law Study*, Larcier 1991, p. 361.

[5] The difficulty in determining the impact of the patent on the drug's price: C. Jordan - Fortier, *Health and International Trade*, preface by E. Loquin, Litec 2006, n° 370.

To fight against the adverse effect of the patent, the solution lies in the development of generic drugs. Indeed, the generic drug is "a legal copy of the original drug patented which patent fell into the public domain" [1], and its price is significantly lower than that of original medicines. Also, governments seek to facilitate the use of these generic drugs to reduce public health insurance spending.

Another criticism became prevalent is the problem of access to medicines. The patent, for the privileged marketing it allows the patentee often makes medications unaffordable for the poorest populations. For this purpose blocking access to medicines, the patent appears as a right toll for access to health care at the expense of fundamental human rights, which is strongly criticized [2]. Also, a differential treatment of medication is offered by some authors [3] on the grounds that access to medicines is a fundamental human right [4]. Without accusing the patent of all evil, it is clear that the pharmaceutical patent system is not working properly, since there is currently a decrease in new developments in the medical field and delayed the entry of generic drugs on the Community market [5]. The analysis of the practices of

---

[1] H. Gaumont - Prat, Law of Industrial Property, Litec 2nd ed. 2009, n° 177; see the definition in the article L. 5121 - 1 5° a) of the Code of Public Health, "a specialty generic reference product, which has the same qualitative and quantitative composition in active substances and the same pharmaceutical form and whose bioequivalence with the reference medicinal product has been demonstrated by appropriate bioavailability studies".

[2] Questioning whether the restriction of access to medicines is not an infringement of human rights: AE. Kahn, *Compulsory Licenses*, in The medicine and person, I. Dupuis (under the direction of), Litec 2007, p. 219, spec. p. 220; see also N. Bronzo, *Intellectual Property and Fundamental Rights*, L'Harmattan, 2007.

[3] Some consider that this property should be public, including for health protection. For a proposal to raise the drug as a global public good using human rights and specifically the right to health: A. Martin, *The Drug, A Commodity Like No Other*, in The drug and the person I. Monk - Dupuis (under the direction of), Litec 2007, p. 279, spec. p. 307.

[4] R. Andorno, *The Human Rights as a Framework for the New Biomedical International Law*, in The medicine and person, supra, p. 311: according to the author, the drugs should not be treated "as mere commercial commodities". About the access to the drug as prerogative of the human person, see also: G. Velasquez, Access to medicines is a human right, but drugs for all are a private matter, p. 117.

[5] The final report of the European Commission referred in particular to the delay of the entry of generic drugs on the market, which is an average of 4 to 7 months: European Commission, Final Report, supra p. 31.

pharmaceutical companies shows that they do not hesitate to divert the patent system to maintain their exclusivity to the detriment of competitors and generic manufacturers[1]. When the patent is no longer enough to ensure exclusivity, then there has been some drift at all costs to maintain the exclusivity of the drug use. The European Commission condemns particularly some of these practices and seems determined to change them. The patent is thus an exercise at the expense of public health marketing tool, as illustrated by the looting of traditional medical knowledge through the patent[2]. Pharmaceutical patent does not sufficiently take into account the public interest and its legitimacy is questioned[3]. Analysis of the relationship between patents and drugs shows that the patent has become an instrument of manipulation used to strengthen the rights of owners, so that the system becomes detrimental to the drug and medical research. This exploitation of the patent demonstrates that the search for a new balance is essential because the evolution of patent law results in an excessive strengthening of the exclusive rights of the patentee.

The evolution of pharmaceutical patent law reflects a strengthening of the monopoly of the patentee. The successive developments of the law of patents have led to the steady expansion of the field of pharmaceutical patent.

B - Criticism of the increase in the scope of the biotech patent

---

[1] To designate this goal blocking or delaying generic competition, the European Commission speaks of "additional" functions to the patent, side of "traditional" functions, matching exclusivity obtained in exchange for the disclosure of the invention: J. Armengaud and E. Berthet - Maillols, Bad use of patent law on pharmaceuticals, according to the preliminary report of the European Commission, Intellectual properties April 2009, n° 31, p. 132, spec. p. 139.

[2] Many drugs are now made from plants that are then patented. But these plants are sometimes used in traditional medicines in developing countries. The exclusivity generated by the patent on such plants can then prevent people from using them and thus deprive them of an existing medicine since immemorial time. This leads to a looting of traditional medical knowledge, and a review is conducted internationally by WIPO to get legal protection for these communities to their knowledge. See in particular: J. Azéma and J - C. Galloux, *International Protection of Traditional Knowledge*, RTDCom. 2004, p. 286.

[3] About the challenge of the "empire" of the patent: M. Vivant, *The Patent System in Question in Patent, Innovation and General Interest*, B. Remiche (under the direction of), Larcier 2007, p. 19.

The scope of the patent has gradually expands to the point where we can say, provocatively, that today almost everything is patentable. And one wonders whether something can still escape the patent [1]. It seems inexorably the patent nibbles gradually extra - commercial spheres when the object has a scientific and economic interest: the patent conquered an ever - expanding field [2]. This progressive expansion of the patent is the result of commodification of objects. The patent does seem to have limits that under specific limits of scientific research [3].

It appears that the patent ensures protection of objects of increasingly numerous and heterogeneous. The principle laid down by the various texts actually remains the possibility of obtaining a patent in any field [4]: there is therefore no general exclusion to patentability. In practice, this ongoing expansion is characterized by an extension of patented in the field of health items, particularly in the biotechnology sector that focus on living and experiencing significant growth through genetic engineering.

The logic of privatization of the patent therefore extends to new objects and leads to a quasi - systematic marketing of products and components from the latest technologies and research. Thus, the distinction between market and non - market values has changed considerably, and there is a constant element once considered non - commercial commodification. As such, the drug now has often living origin: whether the use of animals and plants living raises questions, of course, the difficulty lies in the use of human living and respect for the unavailability of the human body, its components and its products under the article 16 - 1 of the Civil Code. Thanks to ad-

[1] See also Michel Vivant, *The Patent System in Question*, cited above, p. 19.

[2] The issue of software patentability remains controversial and perhaps will know soon clarified by way of the Grand Chamber of Appeal of the European Patent Office seized on this issue: OJ EPO 1/2009, p. 32; about this referral, see also C. the Stanc, Elusive software patent: search view, Prop. Industr. 2009, n° 6.

[3] A. Martin, *The drug, A commodity Like No Other*, supra, p. 301.

[4] According to the article 27 of the TRIPS Agreement, patents may be issued "in all fields of technology" without discrimination as to the field of technology.

vances in biotechnology, medicines are made from human body. By patenting the drug, there is an indirect ownership of the body, its parts and its products. Also, the patent would create, according to some, an unacceptable commercialization of the human person[1]. Because of these ethical difficulties, pharmaceutical products of human origin are being treated differently in different countries, but there's a large admission of the patentability of human living.

The patenting of human genes has long been recognized. Like the inventions in the field of chemistry, the inventions based on biotechnology have been patented progressively on micro - organisms, genetically modified animals and human sourced materials.

The issue of patenting life arose particularly in 1980, in the famous case of Diamond V. Chakrabarty[2]. On the occasion of this case, the question of the patentability of living matter is admitted by the Supreme Court of the United States. With the organization of the human genome in 1990, the sequencing of the human gene pool was implemented and the patenting of human genes has been recognized by many legal systems. The European Directive on the legal protection of biotechnological inventions of 6 July, 1998 affirmed the patentability of biotech drugs[3]. But the transposition of this directive into French law was carried out imperfectly[4], leading the French right to admit very restrictive conditions for the patenting the human body: only the application of a technique based on a genetic sequence can be

---

[1] C. Jordan - Fortier, supra, n° 320.

[2] US Decision 447 303 (1980).

[3] The Directive of the European Parliament and of the Council n° 98/44 of 6 July, 1998 allows in the article 5, 2 patentability of genetic sequences, "an element isolated from the human body or otherwise produced by means of a technical process, including the sequence or partial sequence of a gene, may constitute a patentable invention, even if the structure of that element is identical to that of a natural element".

[4] Expressing a favorable view of the French law on biotechnology opinion: M. Vivant, *Conclusion: The Public Interest Served by an Enlightened Recognition of Intellectual Property in the Public Interest and Access to the Information in Intellectual Property*, Bruylant 2008, p. 277, spec. p. 287: "If the patent is the reward given to the one who makes the contribution of an invention, one must ask what the invention in such cases and in such cases, the invention lies in the development light of a given application".

patented and the scope granted to the patent for a biotechnology is very limited[1]. A clarification of the French law on this point that contravenes EU law is essential and more flexibility also seems desirable[2]. Indeed, restrictive national laws are isolated with respect to this global issue[3], and medical research appears more difficult to carry out in the least permissive states.

Patent law does not consider only the health goods as patentable subject matter: he got to integrate the specificity of drugs and to create specific provisions addressing the needs of innovative companies. The expanding of the field of patent is making it ubiquitous[4] to the point that any discovery seems to be appropriated through the patent. However, the Supreme Court of the United States stopped the excessive potential patentability of biotechnological inventions.

II Redefining the field of biotechnology patents: the refusal by the United States Supreme Court to patent human DNA

By an historic decision of 13 June, 2013[5], the Supreme Court of the United States decides that human DNA can not be patented as such. This solution challenges the boundaries of patentability as they were previously allowed, especially in the United States, and redefines the scope of patentability (A), but the impact of this decision partially terminating the patenting of genes has not been measured (B).

---

[1] Indeed, the article L. 611 - 18, which is the transposition of the article 5 of the Directive by the Law n° 2004 - 800 of August 6, 2004 in the Code of intellectual property provides: "Only an invention that is the application a technique based on an element of the human body can be protected by the patent. This protection covers the element of the human body only to the extent necessary for the implementation and operation of this particular application. This must be exposed concretely and specifically in the patent application."

[2] The next bioethics law will no doubt go further in the admission of the patentability of the human body and on the issue of embryonic stem cells. On the distortion of French law in relation to the Directive and in respect of the European Patent Convention, J. Azéma, Medicine and Patent, Juris - Cl. Patents, fasc. 4280, 2008, n° 9.

[3] About the dangers of the commodification of life: A. Martin, *The Drug, A Commodity Like No Other*, supra, p spec. 303.

[4] About the ubiquity of intellectual property, M. Vivant, *Conclusion: The Public Interest Served by an Enlightened Recognition of Intellectual Property Rights*, supra, p. 277.

[5] http: //www. supremecourt. gov/opinions/12pdf/12 - 398_ 1b7d. pdf.

A – The challenge to the patentability of human genes

Very controversial, the disputed innovation becomes a mere discovery after being qualified as an invention within the meaning of patent law.

There are approximately 20 years, the company Myriad Genetics was the first to sequence the BRCA1 and BRCA2 genes with the University of Utah, genes on which it then obtained patents. However, these genes are responsible for pathologies multiplying the risk of breast or ovarian cancer and inherited mutations greatly increase the risk of developing these cancers. The claims of these patents covered both on isolated mutations on gene sequences but also on any method of detecting these mutations, i. e. on methods of diagnosis of predisposition to these cancers. Thus, due to its very broad patents, the company was the only one able to practice testing of these types of cancers, with very expensive prices.

Since 2009, an important legal battle was fought to challenge the patents held by Myriad Genetics. In support of their action, the various plaintiffs including the Association of Molecular Pathology put forward the need for universal access to these genes so that the disputed patents impede basic research. Besides injury research, the complainants felt that the monopoly of Myriad Genetics prevented the development of other medical tests, thus limiting patient access to prevention. On March 30, 2010, the District Court of New York ruled for plaintiffs : it invalidated the patent on the grounds that the genes were not patentable because the isolated DNA of human genes BRCA1 and BRCA2 was not fundamentally different from DNA origin. On appeal, the US Court of Appeals for the Federal Circuit affirmed the inverse the July 29, 2011 and again the August 16, 2012 by holding that the genes in their isolated state were a product of human intervention.

So the principle of patenting of human genes had been questioned. At the end of the legal battle, the Supreme Court rejected the position of Myriad Genetics. It believes that discovering a molecule of human DNA and isolate is not a patentable invention, but that DNA synthesis can however constitute a patentable creation.

However, the patenting of genes in their isolated state has long been recognized, the isolated genes constituting a product of human intervention. This is what the Court of Appeals for the Federal Circuit decided the July 29, 2011, and then again on the August 16, 2012 in favor of Myriad Genetics [1]. According to this Court, the isolated DNA is a patentable subject matter because the genes in their isolated state do not exist in nature, and are the result of human intervention.

After recalling the Chakrabarty and Funk Brothers decisions [2], the Federal Court held that the isolated DNA molecules have significantly different chemical structures as compared to natural DNA: that identity and chemical structure are unique and that is why they are patentable. Regarding the isolated genes, the Court authorizes the patentability of DNA as it has been modified and does not exist as such in nature. However, the Court refused to allow diagnostic methods are patentable, provided they merely analyze and compare DNA sequences. Therefore the Federal Court reversed that part of the decision of the District Court. The principle of patentability is thus reaffirmed, and the Federal Court did not take into account the Mayo decision of the Supreme Court the March 20, 2012 which dismissed the patentability of a method of blood testing on the grounds that this method adds nothing to the laws of nature and their observation of the fact that it does apply [3]. Also, according to its position, the Supreme Court does reject patentability reaffirmed by the Federal Court in the Myriad Genetics case: changes in favor of the patentability of genes and consecration of broad patents are stopped.

As stated by the Supreme Court, the main discussion was whether isolating DNA constitutes an inventive step to qualify for a patent. The qualification of invention was therefore discussed in the case of having updated the role of a gene in the

---

〔1〕 Federal Circuit, August 16, 2012 n° 2010 - 1406, aff. Association for Molecular Pathology c/ Myriad Genetics Inc. : http: //www. cafc. uscourts. gov/images/stories/opinions - orders/10 - 1406. pdf; H. Gaumont - Prat, USA: patentability of genes, Propr. industr. November 2012, n° 11, 74.

〔2〕 Decision US 333 127 (1948).

〔3〕 Mayo v. Prometheus: Propr. industr. 2012, 58.

occurrence of such disease and had been previously isolated, purified gene and an application that has been proposed.

The previously chosen design was to permit patenting of human genes from the gene has been isolated and described an industrial application was proposed, then there was the invention under patent law.

At the hearing, the attachment of the isolated nature of the state DNA was discussed and unconvincing analogies were made with the example of a harvested in the Amazon for medical plant, the gold that is mined from the ground in its natural state before being traded in the markets, or a baseball bat that is cut in the trunk of the tree. The chosen solution by the Supreme Court is that human DNA is not patentable because it is a product of nature that Myriad Genetics has only discover. Or, to discover is not to invent and can not justify granting a monopoly. In reaching this conclusion, the Court stated that Myriad Genetics has not created or modified the genetic information encoded in the BRCA1 and BRCA2 genes, and that the company has not created or altered the genetic structure of DNA.

Produced naturally by the human being, discovered by biotechnology company genes can not be constitute a patentable invention: the item is not fundamentally different in its characteristics of the element of nature, which was already identified in 2010 by the district Court of New York : it had already invalidated the patents under 35 USC § 101 of the Patent Act[1]. Indeed, the decision had noted that the isolated DNA was not fundamentally different from the original DNA. Genes are only mere discoveries, even if they have previously been isolated, purified, and their application described in an industrial process. The patenting of human genes had been rejected and the decision was criticized because they thought the unquestionable so-

[1] Association for Molecular Pathology V. United States Patent and Trademark Office, n° 09 – cv – 4515, 94 USPQ2d 1683 (SDNY March 29, 2010).

lution and was for some authors dedicated to censorship of the Supreme Court [1]. Thus, human intervention has merely revealed the existence of this object that already existed and no creation has not been performed. Traditionally, patent law excludes from its domain the laws of nature, natural phenomena and abstract ideas, as recalled by the Supreme Court, because it is a "basic tools of scientific and technological work that do not fall within the field of patent protection". Myriad Genetics discovered an important and necessary gene, but it's not possible to get a patent even if it's an innovative and brilliant discovery.

The solution based on the concept of invention patent law is now clear: it is not possible to patent a gene as such it was simply isolated. Technological change probably justifies this change in solution: what was innovative in the 1990's has become an act of observing human nature as a product of nature, human DNA must remain available to all, so as not to affect in particular the interest of public health and society. This rejection of patentability ends a "practical thirty of the American patent office"[2] and breaks the monopoly of Myriad Genetics in the United States. Any patent claiming the genome isolated now belongs to the public domain, and the solution can be extended to isolating any element—cell or bacterium—already present in nature. With this decision, the very threshold of the patentability of living that has been greatly moved, and it is essential to consider the impact that this decision is likely to generate.

B – The impact of the partial end to the privatization of human genes

However, the Court stated that a human gene can be patented in some circumstances, even though it may not be as such.

Patentability is maintained as regards the synthesis of DNA which is copied from

---

[1] C. Noiville and F. Bellivier, Patents on human genes: "Don Quixote to the rescue of public health?", *Review Contracts*, October 1, 2010 n° 4, p. 1417, citing the "courageous questioning of an idealistic judge".

[2] L. Marino, Can you Patent Human Genes?, JCP G n° 29, July 15, 2013, 849.

the DNA of a cell and synthesized artificially. According to the Court, the complementary DNA can be patented because it is not produced naturally. In this case, there is indeed a creation as it is produced artificially by man: a new object appears that did not exist in the state of nature. The invention can be patented and patents held by Myriad therefore remain in force. The monopoly of the company has gone in terms of just single genes but it is supported in the case of the complementary DNA. In addition, tests using genes to detect genetic predispositions are still patentable, but access now allowed to genes will lead other companies to develop tests: Myriad Genetics will not prevent other institutions to develop other tests for BRCA1 and BRCA2.

However, the new frontiers of patentability drawn by the Supreme Court are unclear on what is possible or not to patent. The Court distinguished the natural gene that opposes the gene created in the laboratory, i. e. DNA and cDNA. But from what stage a gene is it not natural? The test used by the Court is the modification of the gene, but we may question the consistency required for the modification of gene which could justify the patent. The solution is likely to cause difficulties in determining the threshold of patentability. Moreover, this decision creates uncertainty about the status of thousands of patents as they relate to a gene or any other elements of the environment simply isolated by man.

Finally, one can wonder about the impact this decision will have on other rights. Does it lead to a change in patent law in other states? In Europe, European patents corresponding to those of the annotated case were finally obtained in a modified[1] following a decision of the chamber of Appeal of the EPO of November 19, 2008[2]. Can European law ignore the decision of the Supreme Court? For example, the article 5. 2 of the Directive n° 98/44/ EC of 6 July 1998 on the legal pro-

---

[1] EP0699754B2, EP705902B2 and EP705903B2.

[2] EPO December n° T 0666/05, 12 – 19 November 2008, www. epo. org.

tection of biotechnological inventions patentable devotes an element isolated from the human body. It contravenes frontally for restraint by stopping commented solution. The French law n° 2004 – 800 of 6 August 2004, which transposed the directive, reduced the scope of patentability to the technical implementation of a function of an element of the human body. So the French text is not consistent with the rejection of patentability of an element isolated from the human body, as it recognizes all the same the existence of an invention relating to the technical implementation of a function of an element of the human body[1]. The decision of the Supreme Court should provoke a debate in states that have also accepted the patentability more or less of the living human. Normative and jurisprudential change is desirable, particularly in Europe, as it is expected to prevent a number of conflicts between the patent first, and secondly research and public health.

In causing the Supreme Court on this issue, the complainants have led to adopt a positive position which will result in significant effects on the patent and in public health. It is regrettable that the judges did not consider the question of the impact of the patent on public health and are not positioned themselves on the issue of obstruction of research and access to care for patients. Indeed, the legal basis of the decision is located previously, upstream, and is to consider that isolating a human gene is not an invention: the discussion does not address the question of the patentability of the human body and the field of patenting life.

The decision will impact primarily on medical research. The monopoly of Myriad Genetics had problematic effect for other researchers, because patents were hard to avoid the obstacles. The barrier to innovation and research that these patents could constitute no more and it should be welcomed. The genes in question may be freely used by research institutions without fear of prosecution, allowing free tests for the benefit of research. The liberalization of genes induced by this decision is crucial for

[1] Art. L. 611 – 18 of the Code of Intellectual Property.

medical research and for patients, especially today: nearly 20% of the human genome is patented, representing 24 000 human genes that are the patented. These patents fall within the scope of the decision of 13 June. This decision may allow us to establish the conditions for a more equitable research.

This decision will then provide more equity for patients to benefit public health. Diagnostic testing from Myriad Genetics are essential for the detection of serious diseases, but their price set between $ 2000 and $ 3000 is three times more expensive than the tests proposed by other agencies. This results in an obvious inequity in access to care. Such testing costs marketed by Myriad Genetics could have led to questions about access for patients to test reasonable prices. However, this very lucrative monopoly will soon be questioned: many businesses and American universities are preparing their tests including the disputed genes and announced their next marketing to a level below that of the test from Myriad Genetics price. Finally, the end of the monopoly will prevent support the additional cost of test by the national social security and health insurance.

The decision of the Supreme Court will reduce some obstacles that patents have contributed to research and access to care: it acts as an instrument of social justice that moderates monopoly and increases access to care. Rebalancing occurs face a monopoly that created imbalances threatening for research, but also for patients. It will take to know the impact of this courageous regulation of intellectual property law by the Supreme Court of the United States.

ETHICAL AND LEGAL FRAMEWORKS FOR EMBRYONIC STEM - CELL BASED RESEARCH IN FRANCE AND IN EUROPE: A CHALLENGE FOR BIOTECHNOLOGY

Anne - Marie Duguet[1], Aurélie Mahalatchimy[2], WU Tao[3], LI Mou[1], Anne Cambon - Thomsen[1], Emmanuelle Rial - Sebbag[1]

Abstract:

Developing research based on the use of biological material raises the question of the sources of the cells. Should we use human stem cells or embryonic stem cells? For human stem cells, since 1994, the French law considers the human body is inviolable and cannot be subject of property (Art 16 - 1 of the Civil Code) that means no trade and no patent. Focusing on embryos, the Oviedo convention (Council of Europe) on Human rights and biomedicine (1997) says in Art 18: it is up to each country to authorize or not the embryo research. There are differences among countries in Europe, from a total ban to an approval of the research with nuclear transfer. In France, several opinions of the French National Ethics committee (CCNE) claimed for the protection of the embryo. Regarding commercialization of cells and patentability, the European Directive 98/44. CE on legal protection of biological inventions says: Biotechnology inventions are essential for the development of the community; only adequate protection can make them profitable and facilitate trade. The

[1] UMR 1027 Inserm/ Paul Sabatier University Toulouse France.

[2] CNRS, International, Comparative and European Laws - DICE - CERIC - Aix - Marseille University, Toulon University, Pau & Pays Adour University; Centre for Global Health Policy, School of Global Studies, University of Sussex, UK.

[3] Xi'an Medical University.

national patent law remains the reference. In France, in 2006 research on embryo and embryonic stem cells was organized by the decree of April 27th, 2006 under the control of the French National biomedicine Agency. Even if the research on embryo was not permitted it could have been conducted under exceptions. In 2010 the French National Ethics committee changed his position in the opinion 112 and the Act 2013 – 715 of August 6, 2013 is now allowing research on embryo and on embryonic stem cells under conditions.

Key words: Embryo; embryonic stem cells; ethics; regulatory framework; France and Europe.

Acknowledgements:

This work is part of the results of the bilateral collaboration between France (INSERM/University Paul Sabatier Toulouse 3) and China (Shandong University – Law School, Jinan), Cai Yuanpei Project (2012 – 2013) n°28007UF. It has been partially supported by the EUcelLEX project (FP7, GA n°601806, PI E. Rial – Sebbag) and by REGenableMED, UK ESRC Project ES/L002779/1, in which Aurélie Mahalatchimy is a member.

Introduction

The use of biological material for biotechnology applications and research raises the question of sources. Should we use human stem cells or embryonic stem cells?

For human stem cells, since 1994, the French law considers the human body is inviolable and cannot be subject of property (Art 16 – 1 of the Civil Code) that means no trade and no patent.

Regarding the use of embryonic stem cells in research, one of the major ethical issues is relying on the destruction of the embryo, the impossibility to ensure its development, and the ethical consequences regarding the principle of respect for the beginning of life.

In France (A), several opinions of the French National Ethics committee (CCNE) claim for the protection of the embryo. From a formal opposition to use embryonic stem cells for research, to an authorization under specific conditions.

In the meanwhile, the research on embryo and stem cells was organized by the decree of April 27th, 2006 under the control of the French National biomedicine Agency. Even if the research on embryo was not allowed it could have been conducted under exceptions. The regime has been modified since the adoption of the Act 2013 – 715 of August 6, 2013, allowing human embryonic stem – cells (hESC) research under conditions.

A quite different evolution can be seen at the European level (B) where the institutions were less keen to frame this issue on research on embryo through hard law instruments.

A – Evolution of the ethical and legal frameworks in France for using human embryonic stem cells in research.

For a long time the ethical issues dominated the debate on the authorization of research on embryos and since 1984, the National Consultative Ethics Committee (CCNE) has produced 8 opinions. The last one in 2010, opinion 112 synthesizes the previous reflection. Meanwhile the legislator in various laws and regulations organized the conditions of access and use of embryonic cells, ending by a change of legal regime in 2013 after deep discussions occurred in the CCNE.

1. The ethical values at a glance: the opinions of the National Consultative Ethics Committee CCNE

The first opinion adopted by the CCNE in 1984 was on "sampling of dead human embryonic and fetal tissue for therapeutic, diagnostic, and scientific purposes".[1] In this opinion the embryo was considered as a potential human being. The opinion followed the first in vitro fertilization was carried out in France. The development of these new techniques questioned the new possibilities to separate reproduc-

[1] CCNE, Opinion n°1 on sampling of dead human embryonic and fetal tissue for therapeutic, diagnostic, and scientific purposes (22/05/1984), http://www.ccne – ethique.fr/en/publications/opinion – sampling – dead – human – embryonic – and – fetal – tissue.html#.VMZJcS5ec4A.

tion from sexuality while also posing challenges to the legislative ability to use tissues that were not previously available for research (in particular for technical reasons).

In Opinion No. 8 (1986)[1] on research on supernumerary embryos, there were large discrepancies among the members of the Committee on the statute of the embryo and on the moral ability to use them (or their cells) for research. Following the progresses on IVF and on the technical possibilities to extract and to store hESC, the Committee supported the possibility to make some research on supernumerary embryos in opinion 53 (1997)[2] and 67 (2001)[3] but only under restricted conditions. Regarding reproductive cloning, in its opinion 54 (1997)[4] the CCNE was clearly against and this position was comforted by the adoption of the Law: Art 16 – 4 al. 3 of the Civil Code stating "Any intervention seeking to raise a child genetically identical to another, living or deceased, is prohibited.".

So, two reactions have developed in parallel, on the one hand the doctrine of the National Ethics Committee on Ethics regarding the possibility of using embryos and embryonic stem cells in research, and secondly a legislative framework to strictly regulate these practices.

The ethical questions which came out from the committee were relating to the

[1] CCNE, Opinion n° 8 on research and use of in vitro human embryos for scientific and medical purposes, (15/12/86), http://www.ccne-ethique.fr/en/publications/opinion-research-and-use-vitro-human-embryos-scientific-and-medical-purposes#.WAYehSRnBek.

[2] CCNE, Opinion n°53 on the establishment of collections of human embryo cells and their use for therapeutic or scientific purposes (11/03/97), http://www.ccne-ethique.fr/en/publications/opinion-establishment-collections-human-embryo-cells-and-their-use-therapeutic-or#.WAYftiRnBek.

[3] CCNE, Opinion n° 67 on the preliminary draft revision of the laws on bioethics (18/01/01), http://www.ccne-ethique.fr/en/publications/opinion-preliminary-draft-revision-laws-bioethics#.WAYgECRnBek.

[4] CCNE, Opinion n° 54 Reply to the President of the French Republic on the subject of reproductive cloning (22/04/97), http://www.ccne-ethique.fr/en/publications/reply-president-french-republic-subject-reproductive-cloning#.WAYg8CRnBek.

nature of the embryo and the cells:[1][2] Is the embryo a person? A bundle of cells? The opinion No. 93 of the Committee adopted on the 13 December, 2006 on the Commercialization of human cell lines and stem cells[3], questioned mainly the ethical values that can be balanced as to the distribution of cells. Two principles were identified and used as a framework for the analysis made by the committee: "the nature and the limits of acceptable commercialization of human cell" and "the parents' consent" to research. Even this opinion is not only covering embryonic stem cells but all types of cells on; the embryo had been largely targeted because of its ontological nature.

After noticing that the majority of the States in the world was considering the human body "out of market", the CCNE admitted, however, that this issue should be reconsidered in light of new practices that tended to develop. Indeed, the human body elements which can currently be completely detached from the body can also be submitted to large transformations and therefore can potentially be turned into products that could be freely commercialized on national territories as goods. At the same time, the French bioethics law applies a common principle to all elements of the body in article 16 - 1 of the Civil Code: "The human body, its elements and products cannot be the subject of proprietary law". This provision totally banned the cells and stem cells from being commercialized in France. Nevertheless, the CCNE applied this principle according a certain degree of flexibility. It insisted on the fact that this principle of non - commercialization should fully apply to the links between the donor and the researcher (no remuneration of the donor, no commercialization),

---

[1] Morris J, Substance ontology cannot determine the moral status of embryos. J Med Philos. 2012 Aug; 37 (4): 331 - 50. doi: 10.1093/jmp/jhs026. Epub 2012 Aug 14.

[2] Lee P, Tollefsen C, George RP. The ontological status of embryos: a reply to Jason Morris. J Med Philos. 2014 Oct; 39 (5): 483 - 504. doi: 10.1093/jmp/jhu031.

[3] CCNE, Opinion n° 93 on the Commercialisation of human stem cells and other cell lines (22/06/06), http: //www.ccne - ethique.fr/en/publications/commercialisation - human - stem - cells - and - other - cell - lines#.WAiROSRnBek.

whereas compensation of costs or commercial uses could apply to "sufficiently" or "radically" transformed elements of the body. These exemptions could also be applicable in case of transformation of embryonic stem cells. The criterion of "transformation" has been discussed in the 93 opinion and has been defined as human manipulations leading to the change of the ontological status of the cells, in other words transform human cell to a cell product. This argument has not been shared by all the members of the Committee as some of them provided divergent thoughts. Indeed, some members argued that the representation of the embryo as laboratory material, underlying the possible commercialization of cell lines, could lead to "crossing a very important line into a trivialization of embryo research". The other issue raised by the commercialization of stem cells, including embryonic stem cells, was relating to patents[1]. The question posed was linked to the nature of the cells which are coming from the human body: should these natural elements be patentable?

2. Research on embryo and on embryonic stem cells: legal overview

In France, research on embryos and hESC is organized by the Decree 2006 – 121[2]. The embryos that may be devoted to research are supernumerary embryos that are no longer subject to a parental project, or embryos not subject to transfer. At the end of the parental project, a prior informed consent specific for embryo research is necessary before starting the research. The written consent of the couple (or of the surviving member), is compulsory after informing on the other possibilities either to destroy the embryo or to make it available for a transfer for another couple. The embryos which are not subject to transfer in utero because of their quality (developmental abnormalities) rejected after preimplantation diagnosis when an abnormality is detected, may be used for research. In all cases, the consent is revocable at any time.

---

[1] Mahalatchimy A., Rial – Sebbag E., Duguet A. M., Taboulet F., Cambon – Thomsen A., The impact of European embryonic stem cell patent decisions on research strategies, Nat Biotechnol., 2015 Jan 9; 33 (1): 41 – 3. doi: 10.1038/nbt.3105.

[2] Decree 2006 – 121 of 27 April, 2006 JORF n°32 du 7 février 2006.

The authorization for conducting research on embryo and embryonic stem cells is delivered by the Biomedicine Agency for 5 years to the establishments holding an authorization for conservation (L. 2151 - 3 - 1 CSP). The biomedicine Agency verifies the feasibility of the protocol, the sustainability of the organization and the research team; the conditions of premises and equipment; the means to ensure the quality, the safety, the traceability of embryonic cells. The Agency holds a national register of embryos and cells (coded).

When research is performed on embryo and embryonic stem cells without respecting the law, punitive sanctions are defined by the criminal code. Sanction for cloning is five years in prison and 7500 000 euros fine (Article 214 - 2 of the Penal Code). The article 511 - 1 PC punishes anyone who lends itself to gamete donation for cloning, even abroad. Regarding embryo research, the Article 511 - 19 PC punishes research without consent and without authorization of seven years' imprisonment.

Until today, there is no jurisprudence of criminal courts for violating the law. Nevertheless there is a decision against an authorization to import stem cells. In 2003 the "Tribunal Administratif" of Paris has received a request from an association defending the "Right to Live" for cancellation of the authorization to import stem cells. The association argued that the research on stem cells is a destruction of the embryo. The Tribunal considered that these stem cells could not be regarded as embryos. The decision was confirmed on appeal by the Administrative Court of appeal[1].

3. Latest evolution of the ethical and legal frameworks in France.

France has undergone profound changes in recent years at the ethical and the legal level. While research on stem cells advances, including embryonic stem cell research, the CCNE doctrine and the legal framework had to evolve to frame the new practices. French law has modified the legal regime for embryonic stem cells research in 2013 as a result of a dual influence: on one hand the ethical assessment

---

[1] CAA Paris 9 mai 2005 (No. 03PA00950).

made by CCNE in 2012 on research on the embryo[1] and the other hand one court decision.

As for the opinion, the committee starts its argumentation by emphasizing on where the majority of States have agreed. "There is one true interdict, shared by everyone: the integrity of embryos in vitro cannot be jeopardized as long as they are included in the concerned couple's plans to—have a child." Insisting, therefore, on parental project as the foundation for the embryo integrity protection (this excludes *de facto* Human Dignity as a principle to ensure embryo protection), the Committee draws consequences to legitimate research on embryos, even though in 2012 we were still under a legal regime of prohibition of research with exceptions. CCNE admitted, according to his previous doctrine, that the "issue of the exact nature of the embryo" is an enigma which cannot be resolved by consensus. The only consensus which can be reached is on the need of a parental project as a source of protection of the embryo, the end of the parental project is considered the only moral condition to make the embryo available for research. Thus, the first condition to be fulfilled to morally allow research on embryos is the expression of consent from the couple to waive their parental project on their spare embryos. This condition obviously assumes that the embryos were originally designed as part of a medically assisted procreation project which thereby prohibits creating embryos for research directly. Thus, as already stated in previous opinions, the Committee legitimates the use of embryos for research purposes in the framework set by the law and since the parental project (which only establishes the link "potential human being") no longer exists. Recall that in 2012 the legal regime prevailing in France was a ban of the research on embryos and embryonic stem cells with exemptions.

This framework has been also modified on a different basis with the adoption of

[1] CCNE, Opinion 112, Ethical reflection concerning research on human embryonic cells and on human embryos in vitro, 21/10/10, http: //www. ccne - ethique. fr/sites/default/files/publications/avis_ 112eng_ 0. pdf.

a court decision in 2012.[1] In this case a Foundation was appealing against a decision of the Biomedicine Agency which was authorizing research using embryonic stem cells. Interpreting the 2011 bioethics law[2] and the requirements for research posed by the law, the Court has considered that two of the conditions required by law were not fulfilled: research should allow "major therapeutic progress" and cannot be carried out "by an alternative method of comparable effectiveness". The Foundation formed an appeal arguing the authorization adopted by the Biomedecine Agency respects neither the one nor the other of these conditions. The Foundation was successful since the Court noted that "the Biomedicine Agency, by merely opposing defense to the lack of scientific consensus on the substitutability of the two techniques, does not establish that an alternative method did not exist at the date of the contested decision." Thus, the judge clearly places the burden of the proof concerning the lack of comparable methods, on the Agency. Indeed, secondly, the lack of demonstration, created ambiguity about the possible "major therapeutic progress", making, therefore, the second condition of the law inoperative.

Since that decision other appeals have been formed but with a limited impact because of the adoption of the law in 2013[3] which now defines new conditions for research in France.

Since 2013 the legal balance has changed in France. Now, research on embryos and human embryonic stem cells is no longer subject to a prohibition regime with exemption but an authorized activity under conditions. One could interpret this legal regime as a move towards liberalization of research[4] but it can also be felt as a re-

---

[1] Cour administrative d'appel de Paris, 3ème chambre, 10 mai 2012, 10PA05827.

[2] Act 2011 - 814 of 7 July, 2011 on Bioethics.

[3] Act 2013 - 715 of August 6, 2013.

[4] Bioy X.; Rial - Sebbag E., L'évolution de la recherche sur l'embryon, une question de principes, Les Petites affiches, La loi. 2013/12; (251): 4 - 12.

sponse to a future market[1].

This change in the regulation framing the research on embryos was also the occasion for the French legislator to soften the legal requirements to realize a research protocol[2]. Since the 2013 law the following conditions have to be demonstrated prior the commencement of the research: the scientific relevance of the research project as well as the medical purpose (this condition is interpreted largely as it includes basic research) ; the law maintains the principle of subsidiarity but in a flexible formulation; ethical principles should be respected but the incentive to promote alternative research is deleted; the consent of the partners is required in the same manner as under the 2011 law, but now parents will not be informed of the nature of the research that will be considered. From this last provision one can say that the law is validating a form of blanked consent from the parents. In fact, as parents are no longer informed on the nature of the future research on the embryos, their consent could be considered valid for any research protocol. This position could have been problematic as it is the first time in the French law where a legal provision legitimates blanked consent. This issue has been partially solved by the adoption of the implementing decree in 2015[3] where it is required from the practitioner to deliver any available information on the research to be conducted in the future. In addition, the new law and implementing decree further specifies procedural requirements for the evaluation, approval and implementation of research protocols.

B – European dimension: ethics and law

For a long time in Europe, ethics committees focused their reflection on embryo

[1] Bioy X. ; Rial – Sebbag E. , L'autorisation de la recherche sur l'embryon: évolution ou révolution? Actualité juridique Droit administratif. 2013/11/11; (38): 2204 – 2208.

[2] Bioy X. ; Rial – Sebbag E. , L'évolution de la recherche sur l'embryon, une question de principes, Les Petites affiches, La loi. 2013/12; (251): 4 – 12.

[3] Décret n° 2015 – 155 du 11 février 2015 relatif à la recherche sur l'embryon et les cellules souches embryonnaires et à la recherche biomédicale en assistance médicale à la procréation [19] https://ec.europa.eu/research/ege/index.cfm.

research, taking into account the different historical and cultural sensitivities.

1. The European Group on Ethics

The European Group on Ethics in Science and New Technologies (EGE)[1] is a neutral forum, independent, pluralist and multidisciplinary; it is composed of fifteen experts appointed by the European Commission for their expertise and personal qualities. The Group's mission is to examine the ethical issues related to Science and New Technologies and on the basis of their work, to submit opinions to the European Commission as part of the development of legislation and the establishment of Community policies

The opinion 9 (1997)[1] focuses on the ethical aspects of cloning and opposes preimplantation into the uterus. Opinion 12 (1998)[2] allows research on embryos within a maximum period of 14 days after fertilization, with parental consent, but forbids the human/ animal hybrids. In 2000 an opinion on stem cell specifics that the egg donation should not increase the constraints on women (Opinion 15,[3] 22).

In 2002, an opinion was issued on the patentability of stem cells.[4] Indeed in 2002, over 2000 patent applications were filed worldwide for human and non-human stem cells among them, 25% for embryonic cells. The patent may cover either process: isolation, enrichment, culture, genetic differentiation modification processing of somatic cells into stem cells; either product: the cells themselves, the lines, the genetically modified cells. The EGE concludes with the need not to ban patent on

---

[1] Opinion n° 9 - 28/05/1997 - Ethical aspects of cloning techniques, http: //ec. europa. eu/archives/bepa/european - group - ethics/docs/opinion9_ en. pdf.

[2] Opinion n° 12 - 23/11/1998 - Ethical aspects of research involving the use of human embryo in the context of the 5th framework programme, http: //ec. europa. eu/archives/bepa/european - group - ethics/docs/avis12_ en. pdf.

[3] Opinion n° 15 - 14/11/2000 - Ethical aspects of human stem cell research and use, http: //ec. europa. eu/archives/bepa/european - group - ethics/docs/avis15_ en. pdf.

[4] Opinion n°16 - 07/05/2002 - Ethical aspects of patenting inventions involving human stem cells, http: //ec. europa. eu/archives/bepa/european - group - ethics/docs/avis16_ en. pdf.

cells or cell lines that would be contrary to the public interest and in the interest of patients. The EGE distinguished according to the nature of stem cell, the lines that have not undergone any changes that do not meet the conditions for patentability, and cells modified by treatment in vitro or by genetic engineering which can fulfill the conditions for patentability (opinion 16, 23).

2. The Council of Europe: The Oviedo Convention

The member States of the Council of Europe, have signed on April 4, 1997 the Convention for the Protection of Human Rights and Dignity of the Human Being with regard to the Application of Biology and Medicine, so called the Oviedo Convention[1]. The convention takes into account all the previous declarations or conventions on human rights (WHO, European social charter, Convention for data protection, rights of the child…).

The preamble recalls the following points. The aim of the Council of Europe is the achievement of a greater unity between its members and that one of the methods by which that aim is to be pursued, is the maintenance and further realization of human rights and fundamental freedoms. The accelerating developments in biology and medicine needs international co-operation so that all humanity may enjoy the benefits of biology and medicine. The Council is conscious that the misuse of biology and medicine may lead to acts endangering human dignity and affirms that progress in biology and medicine should be used for the benefit of present and future generations. The Council considers that measures are necessary to safeguard human dignity and the fundamental rights and freedoms of the individual with regard to the application of biology and medicine.

The general provisions of the convention are the protection of dignity rights and fundamental freedoms. In article 1, parties to this Convention shall protect the digni-

---

[1] Convention for the Protection of Human Rights and Dignity of the Human Being with regard to the Application of Biology and Medicine: Convention on Human Rights and Biomedicine, http://conventions. coe. int/Treaty/en/Treaties/Html/164. htm.

ty and identity of all human beings and guarantee everyone, without discrimination, respect for their integrity and other rights and fundamental freedoms with regard to the application of biology and medicine. Each Party shall take in its internal law the necessary measures to give effect to the provisions of this Convention. The primacy of the human being is recalled in article 2: the interests and welfare of the human being shall prevail over the sole interest of society or science.

Regarding the research on embryos in vitro, the Convention says in article 18 : Where the law allows research on embryos in vitro, it shall ensure adequate protection of the embryo. The creation of human embryos for research purposes is prohibited. We can conclude from this analysis that there is no consensus at the Council of Europe level on the research on embryo. This question is still regulated under national laws and no European framework have been proposed and adopted as a common rule. This position is demonstrating the impossibility, for some very few areas, to reach a European position due to cultural and legal traditions.

3. The legal framework in Europe for research and patent

There are major differences regarding research on the embryo among the EU member states and foreign countries, from a ban in Italy and Germany, to an approval of the research and nuclear transfer in UK, Belgium, Sweden, Canada, USA, Japan, China, Singapore, South Korea. Some countries authorize research, but prohibit the nuclear transfer: Brazil, Australia, France, Denmark, Greece, Finland, Estonia, Latvia, Slovenia and Switzerland.

In the EU each country defines the organization of research as the EU is claiming not to have competency in that field. However, the UE has funded research on stem cells in the 6th and 7th framework research and development program, under the condition of compliance with ethical and legal framework of the country. The EU recognizes the great economic potential of new technologies that is necessary to promote and protect.

Through the Directive 98/44/EC, on biotechnology inventions[1], we can find arguments on the position of the EU on the status of the human body and its elements. The preamble of the Directive states that biotechnology inventions are essential for the development of community. Only adequate protection can make them profitable and facilitate trade. The national patent law remains the reference. The patent is an industrial property that allows especially the limits of use by third parties of the patented invention. To be patentable an invention must: being new, having an inventive step and industrial application

Regarding the patentability of human body elements, the European Directive 98/44. CE says in Article 5: "*The human body ... Including the partial sequence of a gene, cannot constitute patentable inventions...An element isolated from the human body or otherwise, produced by a technical process, may constitute a patentable invention ... the partial application of a sequence or partial sequence of a gene must be described in the patent application*".

Inventions shall be considered non patentable where their commercial exploitation would be contrary to public order or morality; however, exploitation shall not be deemed to be so contrary merely because it is prohibited by law or regulation (article 6). The following, in particular, shall be considered non patentable: processes for cloning human beings; for modifying the germ line genetic identity of human beings; the uses of human embryos for industrial or commercial purposes and processes for modifying the genetic identity of animals which are likely to cause them suffering without any substantial medical benefit to man or animal, and also animals resulting from such processes.

Considering 16 asserts the principle that the human body, at any stage in its formation or development, including germ cells, and the simple discovery of one of its

---

[1] Directive 98/44/EC of the European Parliament and of the Council of 6 July 1998 on the legal protection of biotechnological inventions. Official Journal L 213, http://eur-lex.europa.eu/legal-content/EN/TXT/? uri=CELEX:31998L0044.

elements or one of its products, including the sequence or partial sequence of a human gene, cannot be patented.

There is a European consensus that says: the person to whom the item is taken must have given consent under national law.

In France, transposition of the Directive was delayed because of a degree of reluctance. The matter was referred to the European Court of Justice in the course of an action for failure to fulfil obligations, which led to the ECJ's ruling against France on July 1, 2004, (Aff. C-448/03). "*By not adopting the laws, regulations and administrative provisions necessary to comply with the Directive* 98/44/ *EC of the European Parliament and of the Council of* 6 *July*, 1998 *on the legal protection of biotechnological inventions the French Republic has failed to fulfill its obligations under Article* 15 *of that Directive. The French Republic shall pay the costs.*"

The provisions of the Law on Bioethics voted on August 6, 2004 as a result of the transposition of Directive 98/44 were a compromise solution and attempted to temper the scope of patents. CPI Article L. 611-18 transposes differently Article 5 of the Directive by stipulating that "only an invention constituting the technical application of a function of an element of the human body can be patent-protected". Nevertheless, significant divergence between the formulation of the above text (and that of CPI Article L. 613-2-1) and Article 5 of the Directive dated July 6, 1998, is problematic.

In conclusion, the French law has evolved in a sense to open the conditions for conducting research on embryos and stem cells in France. This position was the one claimed by the French researchers notably because of their willing to contribute to the international "race" which the previous legislation was impairing. However, even though we can support this vision, the French law and its interpretation will have to ensure a high level of protection of the parents and of the embryo, opening the research should not have to be detrimental to the respect of fundamental rights. At the European level, we can conclude that the European institutions do not want to go further on solving the issue from a regulatory point of view. As it is considered as a very problematic issue, only the European Group on Ethics has decided to propose some values and rules about embryo research. In a sense, these values can be considered as common roots for considering embryo research acceptable in Europe.

HUMAN DIGNITY IN THE LEGAL REGULATION OF HUMAN GENE TECHNOLOGY INTRODUCTION

MENG Wen[1]

With the assistance of technology, human being has greatly increased the ability of knowing the world beyond natural limitations and weakness. Since the Industrial Revolution, many natural elements have been applied into life and production such as wind, rain and thunder. As nature is no longer a purely mysterious existence but a treasure house of resources serving for human, we expanded our naked eye vision by telescope, extended information access and operation ability by computer technology and enriched our perception by radar technique. As the human's feeler for changing the world shifting from the external to internal, Human Gene Technology appeared, becoming an important technique that may dominate our future.

Human Gene Technology is the term given to a range of activities concerned with understanding gene expression, taking advantage of natural genetic variation, modifying genes and transferring genes to new hosts[2]. The main part of Human Gene Technology includes gene therapy[3], gene screening[4], genetic

---

[1] office worker of school of management of state Grid, Doctor of law (Shandong University), PhD of Biological and Medical law (University Panol Sabatier).

[2] http: //www. csiro. au/Outcomes/Food - and - Agriculture/Gene - technology. aspx.

[3] Gene therapy is the therapeutic delivery of nucleic acid polymers into a patient's cell as a drug to treat disease.

[4] Gene screening is the quick evaluation and acknowledgment of the biological function of new gene and ensure the gene whether could be the research of drug targets or therapeutics (genetic or protein). Reviewed by Yin Hong& Du Guanhua, "The Progress of Screening Methods of Functional Genes", *Foreign Medical Sciences Section on Pharmacy*, XXX (Dec. 2003), p. 321.

testing[1], stem cell research[2] and cloning[3], etc. All the techniques above are on the basis of cognition and correction of human individual gene, the research is focusing on improvement of human health conditions and controlling life mysteries by grasping and modifying the human cell. In the process of exploring the legal regulation of human gene technology, human dignity, as the purpose of the legislation, was repeatedly pushed and strengthened, becoming the basis of identifying fundamental rights such as the right to life, the right to health, the right to physical integrity, privacy, informed consent right and patent right in human gene technology. Carding and analyzing the concept of human dignity in the legislation can help to understand the importance and necessity that the human gene technology challenges still should adhere to human dignity as the legislative purpose when human dignity is under great controversy. And help to grasp the balance of value in the legal regulation of human gene technology.

The dispute of human dignity in the legal regulation of human gene technology

The world "Human Dignity" itself originates from western religious view, its root can be traced back to the "Judaism" doctrine of Hebrew culture in the year of

---

[1] Genetic testing, also known as DNA testing, allows the genetic diagnosis of vulnerabilities to inherited diseases. In addition to studying chromosomes to the level of individual genes, genetic testing in a broader sense includes biochemical tests for the possible presence of genetic disease, or mutant forms of genes associated with increased risk of developing genetic disorders. The results of a genetic test can confirm or rule out a suspected genetic condition or help determine a person' chance of developing or passing on a genetic disorder. http://en.wikipedia.org/wiki/Genetic_ testing.

[2] Stem cell research is mainly applied into gasping the human body cells for curing disease. Stem cells are undifferentiated biological cells that can differentiate into specialized cells and can divide (through mitosis) to produce more stem cells. In humans, there are two broad types of stem cells: embryonic stem cells, which are isolated from the inner cell mass of blastocysts, and adult stem cells, which are gound in various tissues. http://en.wikipedia.org/wiki/Stem_ cell.

[3] The word clone has two meanings, one is a group with the same genetic background, the other is the process of producing similar populations of genetically identical individuals that occurs in nature when organisms such as bacteria, insects or plants reproduce asexually. Qiu Xingxiang et al, "Bioethical Argument by Cloning Technique", *Chinese Bulletin of Life Science* No. 11 (2012), pp. 1302 - 1303.

second century A. D, man as the creatures of God with no distinctions are created equal, and the life given by God is full of sanctity, deserving value and protection[1]. As the same situation of "Human right", when the concept of "Human dignity" first leaped out at human, what brought not only individual liberation excitement and passion, the regression of rationalism and throbbing, but also a process of conceptual globalization and collision with different cultures. In the face of problems caused by human gene technology, though many scholars claimed that this concept is over of vacuousness, lacking in practical operability, this concept as the legislative purpose of human gene technology will make contribution for people to study the primary objective of human gene technology research. "Human dignity" itself has great diversity with sources of different levels which brings different value judgment for it. Thus, the elaboration to human dignity is always full of contradictions and critical comments without a final disposal. Only to clarify the human dignity on different levels can we really understand the reason why to take human dignity as the legislative purpose of human gene technology.

The religious perspective

Human dignity is the core value of Christianity and Catholicism to the religion[2]. Catholic doctrine insists: "human dignity is rooted in his or her creation of God's image and likeness." The Church said that everyone must have the same image as the God's creation, with personal dignity. Human dignity is also an important element in Judaism doctrine. Offending the dignity of the recipient is avoided by warning the public charity in Jewish Law. In Islamic culture, it is made clear: "True faith is to ensure strengthening dignity forward along with the road of human perfection (basic human rights)"[3]. These religious elements brought peo-

---

[1] Long Sheng, "The History of Human Dignity in Legal Concept", *The Rule of Law Forum*, (Jul. 2008), p. 58.

[2] O'Hara, Phillip Anthony, *Human Dignity, Encyclopedia of Political Economy*, 1999, p. 471.

[3] Mayer, Ann Elizabeth, *Islam and Human Rights: Tradition and Politics*, Westview Press, 2006, p. 62.

ple a dispute about the principles of human dignity, especially a religious classic, The Bible. In the Bible, man is created in God's image. Therefore, man has a lot of similarities with God, has the innate, non – transferable and inseparable dignity[1], which gives man right to control with God and makes the man have the ability to understand the whole and comprehend its advantages in the whole independently. From this point of view, the responsibility of human to continuously perfect ego can be concluded, in order to complete the creation of God. Wherefore, starting from the perspective of human dignity, not only did some people support the general treatment and medical practice, but believe some defensive behaviors such as in vitro fertilization or clone are natural restorations on the road to imitate God, that makes restriction of human gene technology through legislation meaningless. However, if people understand and accept in violation of nature will be punished by the concept of human dignity based on the Bible, then the instruction of the Bible may change to another direction—humble: although man was created in God's image, the mankind itself is not divine; humans are creatures rather than creators. Seen in this light, whether body in sickness or in health, heart in purity and integrity or in physical and mental deformity, human all enjoy the God given dignity. Dignity given ethical guidance in the level of humility, explains for human even tiny embryos what should have in the first and the last of life regardless of disability or dementia. Therefore, human only evaluate the value of others in accordance with the way of evaluating required by God, without creativity beyond the requirements of God.

So the human dignity is based on religion, on one hand, some scholars believe the human dignity which are led out with it has the innate fallibility, just religious dogma camouflage in front of the reality of bioethics without legal status. This makes some fictional mythical factors on respecting for human beings and human autonomy,

---

[1] Leon R. Kass, *The Beginning of Wisdom: Reading Genesis*, New York: The Free Press, 2003, pp. 36 –40.

and takes the opposition of human gene technology which may change human genetic inheritance as a saint sanction by the order of nature. On the other hand, some scholars believe that it is of great value to guide the legislation of human gene technology from religious perspective, which can give a hand to (regardless of whether we were followers) explore the most fundamental intuition, the distinctive power and behavior of human being, besides, firmly convinced of the rights and responsibility that man have. Moreover, from a purely secular perspective, it will bring greater difficulties for the popularization of human dignity to close all religious reasons which are known with dignity. Overall, the human dignity on religious aspect clearly indicated that human gene technology is anti - natural. Opposition to abortion, opposition to research embryos and opposition to cloning technology prevent human from becoming God and becoming the theme of human dignity under the religious meaning.

The philosophical perspective

As early as in ancient Greece and ancient Rome School of Philosophy, the Stoicism insists that dignity as a real existence takes in all mankind no matter of their personal circumstance, social status or working attainments. The truth and universality of dignity is due to the rationality and best life that human have, this life is from nature, which helps everyone choose after deliberation. At the same time, human can grasp the personal happiness and inner peace under the domination of rationality. Despite of the existence of poverty, disease or oppression, there still are ways of living in dignity, with integrity and dignity can be deprived by no one. To the Stoics, dignity as a profound idea of democracy, not only can exist in the vulgar and in the noble as well, is applicable both slaves and the king[1]. In the Renaissance, Italian philosopher Pico Mirandola[2] emphasized from the angle of "man is created

---

[1] Adam Schulman, *Bioethics and the Question of Human Dignity*, *Essays Commissioned by the President's Council on Bioethics*, Human Dignity and Bioethics, Washington, D. C., March 2008, p. 7.

[2] Pico Mirandola, *On Human Dignity* (Beijing: Peking University Press, 2010), p. 25.

by God" in his book On Human Dignity that man as the center of the world is free, "a creature of undermined image". In his book, the creator of Adam said: "You are not subject to any constraints that limit the freedom to choose your nature following your free decisions; we have given you to your free will. We have placed you in the center of the world, where you have more possibilities to stare at everything. ... you are the honorable and free self - dominator." [1] Thus, man has the endowed freedom, "he was allowed to get what he chose and become what he want", If he is not satisfied with the fates of any other creation, he will resettle in the center of self - synthesis, becoming the sole spirit with God, thus, "Once is placed above everything he will surpass all" [2]. Obviously, as the word of human dignity is firstly put forward as a philosophical word, it is on behalf of the supremacy of human, the innate freedom and individual autonomy.

It is Kant who made human dignity of philosophical meaning develop to the highest point. And he is called "the father of the concept of modern human dignity" [3], in his view, human dignity is no longer respected based on religious concept [4]. Two aspects are included in Kant's theory of human dignity, one is human nature, dignity is the other [5]. In Kant's time, naturalists believe: "They may break from philosophy only through ignoring and insulting it. But, they can not move a step forward without thinking, besides, thinking is always companied with logical categories...Thus, they became a total slave of philosophy, unfortunately, most of

---

[1] Pico Mirandola, *On Human Dignity* (Beijing: Peking University Press, 2010), p. 25.

[2] Pico Mirandola, *On Human Dignity* (Beijing: Peking University Press, 2010), p. 28.

[3] Bognetti, "The Concept of Human Dignity in European and U. S. Constitutionalism", in G. Nolte (ed.), *European and US Constitutionalism, Science and Technique of Democracy* No. 37, 2005, pp. 75, 79.

[4] Christopher McCrudden, "Human Dignity and Judicial Interpretation of Human Rights", *The European Journal of International Law* Vol. 19 no. 4, 2008, p. 659.

[5] The President's Council on Bioethics, Human Dignity and Bioethics: Essays Commissioned by the President's Council on Bioethics, Washington, D. C. March 2008. https://bioethicsarchive.georgetown.edu/pcbe/reports/human_dignity/chapter13.html.

them are the worst one." [1] And aware of the limitations of the perceptual and the rational, Kant's goal is to improve the metaphysics to the equal position of science and mathematics, making the philosophy free from the subjective impression of people.

In order to achieve this goal, Kant demonstrated the importance of human dignity through a gradual process. Firstly, he was concerned with human itself through the understanding of knowledge and freedom, laying emphasis on the values and meaning of human being. He distinguished the inner world and the outer world through the space definition of necessary requisite to the external sense; and established the sequential quality of human behavior through the pure framework of time of the internal sense; developed a mode of analytic demonstration which can be founded when reason is beyond its legitimate areas. "I call all knowledge transcendental if it is occupied, not with objects, but with the way that we can possibly know objects even before we experience them". [2] The transcendental implies the inducible prerequisite and possibilities to prompt something to happen. The purport of transcendental aesthetics is that, humanity alone, is the fundamental aethesia of extrinsic time and space. Next, Kant got access to the concept of "God" through the ethics, reasoning that there must be something in human nature causing or leading ideas with full acceptance. Meanwhile, the cognition of human being should be dual, that man are the citizens of two worlds, the natural and the noumenal. People' behavior is instinctive and obliged in natural world, but in world of morality and ethics, people should obey the rule with principles of right and wrong. Man can not control the natural attribute, but his behavior will be responsible for his chosen principle when decide to do something. In the existentialism theory of French philosopher, Sutter, man is free because we always own the possibility. That the moral law is uncondi-

---

[1] Friedrich Engels, *Dialectics of the Nature* (Beijing: Peoples Press), 1971, p. 137.

[2] Kant, *Critique of Pure Reason* (Beijing: Peoples Press), 2004, p. 19.

tional is combined with the characteristics of human practice principle, a opinion becoming into existence that, one of the features which makes us human is unconditional, freedom of the transcendental. It's a resolution of inner heart with inwardness not influenced by any material wants. At last, Kant proposed the moral philosophy, the deontology, some scholars also called it as "absolute command", which is the core of human dignity. Among which, the voluntary formula [1] is the most important, strengthening that it is important for the agent to convince himself as the legislator, to independently choose and obey the law he made. Another important judgment of deontology is that one thing is valuable or considered to be good, or it is the means to realize a worthy goal, or it is good itself. The first half of this representation is the good of means and the second half is of the purpose. In Kant's opinion, when one thing (or event) 's good is "the good of purpose", if and only if, the good of it does not depend on any good of other things and it is still good even if it was separated from other things. When one thing's good is "the good of means", if and only if, the good of it depends on "the good of purpose", in another word, it is the mean to realize "the good of purpose" directly or indirectly. According to Kantian perspective: "You should conduct your actions that, the human natures of your moral quality or of someone else, would not be considered as instrument if could be as purpose." [2] "The act is to be morally right if and only if the actor would take no one as the instruments". [3] In order to achieve this, the dignity of human deserve to be respected. "A person with reason also is a purpose based on his nature, thus, he is the purpose of himself and as a code, he must be a criteria to restrict all the purpose

---

[1] One's act is to be morally right if and only if the actor consciously abide by the legal principles of the act; the actors abide the principles mainly because his thinking of themselves as legislator.

[2] Kant, *Immanuel Kant—Complete Works*, Vol. 4 (eds. Li Qiuling, Beijing: China Renmin University Press, 2005), p. 25.

[3] Fred Feldman, *Introductory Ethics*, N. J.: Prentice - Hall, 1978, p. 120.

of the merely relative and with subjective arbitrary". [1] Only when human are endowed with personality, the subject of moral practical rationality, will surpass all prices; because as this kind of people (as the noumenal), he should not merely be evaluated as the means to achieve others' ends even the means to achieve his own purpose but should be evaluated as ends in themselves, that is to say, he owns a kind of dignity (an absolute inner value), whereby, he enforces other existent with rationality in this world to look up to him, comparing with others and evaluating himself on an equal basis. [2] Kant laid emphasis on pure self – determination in his human dignity theory, which means man should make self – decisions with dignity. Self – willing is the basis of human dignity. On one hand, there exists an causal sequence in selfwilling which is not only under the influence of natural regulation; on the other hand, selfwilling means man can take control of their own. [3] At the same time, Kant's concept of human dignity highlights the purpose but the means, which became one of the basic starting point of scientific legislation.

However, many scholars suppose there exist some problems on the application of Kant's concept of human dignity in human gene technology. On one hand, Kant made a breakthrough in the reconciliation of morality and mathematical physics: human dignity is completely defined from the perspective of autonomy with sense, for realizing this goal, he had to reject other aspects of human nature which is of great moral importance, including our family life, our love, fidelity and other emotions, as well as the way we came to this world and other biological facts about human body functioning. But these made the cognition of human moral life too parochial and lim-

---

[1] Immanuel Kant, *Groundwork of the Metaphysics of Morals*, translated by H. J. Patom, New York: HarperCollins, 1964, p. 104.

[2] Kant, *Immanuel Kant—Complete Works*, Vol. 6 (eds. Li Qiuling, Beijing: China Renmin University Press, 2007), p. 445.

[3] Thomas Christiano, "Two Conceptions of the Dignity of Persons", edited by B. Sharon Byrd and Jan C. Joerden, jahrbuch fur recht und ethik [Annual Review of Law and Ethics], Berlin: Verlag Duncker & Humblot, 2008, p. 4.

ited, and arouse many controversies to discuss human gene technology from such small area. If the issue of human dignity can be settled only by rational autonomy, the objections to cloning of human embryos, building muscles by drug help and emotion - controlling will not violate the conditions of human dignity. On another aspect, individual rational autonomy is theoretically lucid and without any disputes, but it is hard to undergo in practice, especially in bio - medical field. For example, if dignity depends on rational will, whether the patients who lost their wills because of disease have dignity. On the third aspect, the moral philosophy of Kant has left a lamentable residual to the later ethics thought on the respect of strict distinction between deontology and consequentialism, between absolute order of morality (Kant's opinion) and supposing that our behavior is only with good or bad consequence (the opinion of utilitarianism). The bioethics required by human gene technology in practice calls for a proper balance of ancient wisdom and it is inadvisable to dogmatically insist on the supposed division between ethical process and ethical results. [1] As for the human dignity on pure philosophical perspective, it is filled with uncertainties, and philosophers has made more specific complements to the concept of human dignity based on the theory of Kant. The requirements of protecting human dignity is not only protecting individuals from infringement, but the state to assist each other to achieve and maintain happiness. [2] Moreover, to protect human dignity need to realize the equality between people, equality of resource and equality of opportunity are involved. And many scholars believe that the theory of human dignity should be applied under the political background. As John Rawls thinks human dignity itself can not play the basic role: the content of this concept can not be obtained without cer-

---

[1] Adam Schulman, "Bioethics and the Question of Human Dignity", *Essays Commissioned by the President's Council on Bioethics*, *Human Dignity and Bioethics*, Washington, D. C., March 2008, pp. 10 - 12.

[2] White Mark D., Dignity, Jan Peil, *Handbook of Economics and Ethics*, Edward Elgar Publishing, p. 85.

tain political principles. [1] Dworkin defined the two principles of human dignity from the perspective of social systems construction. One is about the intrinsic value of the principle that there is a special objective value in everyone's life with innate importance. But this is a kind of potential value that needs everyone's endeavor or it will be dilapidated. The other is about personal responsibility that everyone assumes a special duty to achieve self – success, which includes the pursuit of the success defined by oneself. And this is the responsibility on a level of life sovereignty and the criteria is free and to immune the influence of other people. The two principles established the basis and conditions of human dignity. And the principles are not completely within the category of individualism, the social meaning is also involved. That indicates equality and freedom in the aspects of social practice. [2] In the theory of human dignity of Dworkin, the principle of human dignity is not merely a castle in the air, on one hand, the intrinsic value of human should be upheld and aggressively pursued, to keep the equality between people from this perspective; on the other hand, it will embody the free will in the life choice of oneself, and respect for the choice within the informed reasonable limits.

The legal Perspective

The trend that the protection of human dignity is embodied in laws of nations was started from the end of the Second World War. From the original text, all the theory relied upon is to like the root of American Revolution (life, freedom and happiness pursuit), or the "life, freedom and property" in the Locke's classics, as well as other theories of modern natural rights theorists. There are many countries whose

---

[1] John Rawls, *A Theory of Justice*, Cambridge, Massachusetts: Harvard University Press, 1971, p. 586.

[2] Ronald Dworkin, *Is Democracy Possible Here?: Principles for a New Political Debate*, The United Kingdom: Princeton University Press, pp. 9 – 12.

constitutions guard the human dignity, like German[1], Iran[2], and Switzerland[3]. At the same time, France[4] is a typical country whose civil law protects it. The law which refers to medical ethics, such as in many international documents, The Nuremberg Code, Declaration of Helsinki, and in domestic legislation of many countries like France and Canada, all clearly identified that human dignity - orientation shall be the legislative purpose to guide the research related to medical ethics. At the current period, human gene technology still belongs to medical area, although attracting more and more attention with the trend to become a new field of study, it is more applicable to medical ethics in the law. In some legal documents required by specific technology, like in the declaration Human Genome Project, human dignity is frequently mentioned as the purpose of legislation.

However, some scholars believe the application of human dignity as the supreme principle in the laws indicates that the rights and duties of man rest with a supreme value, with still no clear definition of meaning, content and foundation[5]. In the current human dignity - related legislation, human dignity still is an unknown concept with no specifically defined. But the other scholars suppose the contents of human dignity requires simple specification and expansion with the connotation of the principle of voluntariness and informed consent; the need of protecting confidentiali-

---

[1] German protects human dignity on the first article of the Constitution, as the basic principle of German Constitution: "Human dignity is inviolable. To respect and protect it is the duty of all state authority."

[2] The need to respect human dignity has been written in the Iranian constitution law. Article 2 of the Iranian Constitution Law mentions six principles and infrastructures as basic to the governing system which in Article 1 is called the Islamic Republic of Iran. The sixth principle of this Article concerns human dignity and stipulates that "the Islamic Republic of Iran is a system founded on faith in.... 6) Human dignity and high value and his/her freedom as well as his responsibility before God".

[3] The Swiss Federal Constitution provides in article 7 that "Human dignity must be respected and protected".

[4] The French Civil Code provides in article 16 that "legislation ensures the primacy of the person, prohibits any infringement of the latter's dignity and safeguards the respect of the human being from the outside of life".

[5] Shultziner, "Human Dignity—Functions and Meanings", *Global Jurist*, Topics 3, 2003, pp. 1 - 21.

ty; prohibition of discrimination and abuse[1]. No matter now or future, the possible disputed questions that may occurred in the relevant legislation of human gene technology can be solved within the frame of these principles.

Through different levels of disputed analysis to human dignity, it is not hard to see the concept of human dignity no matter in religion or in philosophy has influenced human's understanding of it. Taken together, the main problem the concept of human dignity in classical sense brought to the relevant ethical controversy and legislation of human gene technology is out of its ambiguous relationship with technical progress, aristocracy status and inequality orientation, of which, the application of the concept of human dignity of the Stoics in relevant areas of human gene technology is limited because of his exacting standards and indifference to the outer world including the bodily sufferings. Although the Bible instructs the human dignity is ample and full of emotions, it also has the possibility of causing ethical and legal confuse of human gene technology. The reason is that it not only indicated the God - like control of nature but the sacred humble recognition of humans' lives in all forms as well. Both the moral philosophy of Kant and the current legislation of the principles of human dignity in gene technology, support and protect the belief that all human beings enjoy the same dignity. But the theoretical structure of Kant's human dignity has confined its application in the legislation of human gene technology. But the specific regulations in relevant legislation failed to invoke the specifications on the content and background of human dignity, which manifested the conceptual excess of the phrase. On the other hand, how to maintain the ethical standards of recreating human gene as it is hard to figure out what's the due without real concept and nature of human beings. Those ambiguous explanations are the origins of disputation in human gene technology legislation. Although the origins are shown to be able to induce

[1] Ruth Macklin, "Dignity is a useless concept", BMJ 327, 2003, pp. 1419 - 1420, http://www.bmj.com/content/327/7429/1419? etoc., the lasted visited on May 12, 2014.

human's doubts in the concept of human dignity, human still are willing to stick up human dignity, for in the history of western political philosophy, no matter Hobbes, Locke or the funding fathers of USA tried to protect the right and freedom of human out of political agenda with the respect of human dignity. The thoughtful and deliberate requirements to the contents and basis of human dignity is for better service of liberal democracy, promoting lenient, freedom, equality and peace. Meanwhile, it can not be denied that some principles derived from human dignity, such as the principle of voluntariness, informed consent, confidentiality, anti - discrimination and anti - maltreat have played a positive role in the protection of human gene technology for a long time.

The necessity of sticking to human dignity in the legislative regulation of human gene technology

In the legislation of human gene technology, human dignity plays a role of ensurance and guarantee for the value trade - off function of the law. As a main regime of adjusting social relations, law is gradually realizing this function. As Pound remarked: "In all the classic period of the law history, both in ancient and modern world, the reasoning of values, critique and logical applications, were the dominant activity of jurists. "[1] As a new high technology, human gene technology, the basis of law to regulate it lies in the balance of values and the basis of law to balance the values is human dignity, reflecting the necessity of upholding human dignity in the legislation of human gene technology.

The cornerstone retaining human nature

Human dignity comprises some substantive parts, such as human supremacy, people's self - determination, human subjectivity, human equality, personality, no discrimination, on harm and so on. The reason why human dignity is the legislative purpose mainly contributes to the cognition of human nature in this concept. First of

---

[1] Ezra Pound, *Social Control Through Law* (Beijing: The Commercial Press), 1984, p. 55.

all, in the human dignity, human is the product of nature, the natural attribute is the foundation of existence. From a biological perspective, all individuals have the common traits in the body, organ, cell structure regardless of nationality, area or race. And because the traits can be inherited without the limitation of human, the evolution of human is of naturalness. Although human is the advanced creature of nature with the ability to walk upright, manipulate a tool and use hands, human is incapable of divorcing from the attribute that nature gave however not restricted the human's moral world is from the human's body, which makes human keep the nature during the process of evolution, living and reproduction. What is not comply with the nature is not complied with the human essence. Therefore, the human gene technology which may break human essence doesn't comply with the nature and the requirements of human dignity. From the whole perspective of human being, the foundation of the overall constructional means of today's society is "natural person", reproducing by natural way—sexual reproduction, and produce natural individuals. Whether the base of western Liberalism—the sate of nature, nor the base of eastern Family Philosophy—natural family can challenge the cornerstone, "natural person". Thus, any system of government and system of state shall not violate the rule: "natural person" forms the "natural family" by marriage, "natural families" form the society and to safeguard the "natural person" and "natural rights" established the country. Before the appearance of an theory and practice of better breaking the social system based on "natural person", in violation of the natural attribute of "person", the legislation will lead to the social disruption and incapability of action.

Secondly, in human dignity, human, by nature, has the possibility to transcend the nature. As Pico said in his "On Human Dignity": "God created man but didn't give him a specific role, which makes people have their own positions." This makes people have the moderate possibility to get rid of the natural world and legislate for themselves. As early as in the Aristotle's time, the mental and rational competence of human which gave human an intelligible world independent of the natural world had been involved in the difference between human and other animals. In Kant's opinion, people become the citizens of two worlds—one as a natural creature, living in the "perceptual world", shall obey the determinism of the world; another, as a kind of life with thoughts, belonging to the "thinking world", is free as well[1]. Here, man is not only an unity with subjective initiative and the ability to think independently which differ from other animals, but also an unity to be his own master in the separate innate world as well, which makes the self - development of human exist fortuity and break the developmental orbit of the law of nature. And just because man can position themselves, their identities are different and they can arrange their lives according to their preference, which gives people various characters.

Thirdly, the universality and particularity in human dignity is a relative sense, man is a creature living in two worlds, one is the natural world and rational world is another, and man is free to make decisions between the two worlds. No matter how will the science and technology develop, as a member of natural world, human can not come out of thin air even cloning is also based on the existing human cells or genes. Notwithstanding human has the competency to use tools, the objects for transforming still should be drawn from the nature. In the whole ecological environment, human not only plays a role at the top of the food chain, but also takes the responsibility of making decisions for the ecodevelopment. Those roles and responsibilities may not answer human's will. But, with the increasing requirements of the quality of life, increasing population and accumulation of knowledge and so on various factors, human's demand of nature will increase. However, the nature will change and wherever the object for human's transformation will extend, the choosing right must in the hands of man under the current framework of the world. So it is at this point that "man" should have the absolute right to self - determination which entitles to every natural person. In this way, the stability of current social systems can be guaran-

---

[1] Kurt Bayertz, *Gene Ethics*, Beijing: Hua Xia Publishing House, 2000, p. 109.

teed.

At last, in the sense of human dignity, "man" is the fundamental starting point of the Eastern and Western Philosophy and is also the fundamental position in law concept that man is the center of the country, society even the world, no matter the scientific technology or policies and laws, ensure the survival and development of the human – centered. In the case of the west, early religions record that human were endowed special rights by God or the Creator which is different from other animals. Starting from the ancient Creek period, westerners were keeping forward on the exploration of human's nature. Western philosophy brought rationality to the world of human without any religious factors. So far, rationality became the root of the human – centered and a set of social, political, economic, and legal system are established with rationality as the core. In China, it was "Harmony of Heaven and Man" that is one of the cores of ancient philosophy, and "human" is the most valuable part of the world. Although there was no good solution to the equality among people, the importance of human as the center of the society was by no means less proficient than western world. The architectural approach of state policies which is centered on human, will give man subjectivity in every aspect of human's life including law, ethics, philosophy, politics and technology. The subjectivity and centrality of human is on the top priority in human dignity.

It is likely to say that the development of the human gene technology leads to the nature quality and rational quality transferring from a hypothesis to a issue of actuality and emerged contradictions. The present legal system centered on human reason demanded that the "man" of dignity under the embodiment of international society is inevitably defined and completed conforming the aspect of nature, on the premise that the subjectivity, centrality, personality, self – determination of people are always in continuation and the guaranty from the ancient time. In the social system of no circumstance of accommodation to the "unnatural", only the "natural" can protect the issue of society and the dignity of human life.

Holding the bulwark of freedom

Science should be at its utmost freest without any disturbance of anyone and scientists can freely probe into the research object to achieve the regular understanding and mastery of it. National constitutions are enacted to protect freedom of science as a basic civil right. Article 47 of the constitution stipulates: "Citizens of the People's Republic of China have the freedom to engage in scientific research, literary and artistic creation and other cultural pursuits. The state encourages and assists creative endeavors conducive to the interests of the people that are made by citizens engaged in education, science, technology, literature, art and other cultural work." There is no doubt that human gene technology is within the range of freedom of science. "The Statements of Embryonic Stem Cells" of Human Genome Organization has mentioned explicitly that "the freedom of research should be promoted" and "scientists, as the bolster of the development of knowledge and understanding, deserve supporter and insistent. We should encourage creativity and innovation, accordingly, the intellectual property rights of researchers should be reserved. The tensions between individual rights of scientists and the legitimacy of the general statement that gene information are used to promote the maximum benefit of all man kind should be acknowledged and solved." As Kaufman suggested in his Legal Philosophy: "We must keep open - minded to accept new ideas in order to handle the future challenge"[1].

However, free is limited and the limits are set by human dignity. Talking of the freedom of human gene technology under the dimension of human dignity, the self - determination is a key point. The individuality and sociability of people cause the individual self - determination to face the problem of choosing between individual rationality and collective rationality. This problem is hard to get a solution from the ethical perspective, because as a member of society, it is reasonable for safeguarding the human common heritage and existential security to forbid from modifying human

---

[1] Kaufman, *Legal Philosophy*, Beijing: Law Press China, 2004, p. 438.

gene but there still exist possibility of disagreeing with the freedom principles required by individual rationality. Meanwhile, this problem also invites another question about the attribution of children: for if the children are attributed to society, the eugenics will have the meaning of optimizing allocation of resources, meeting the requirements of group rationality and hard to ban it theoretically; and if the children are attributed to their parents, the basis is even less to forbid parents from prenatal screening based on any reasons. But these contradicts can not shake the positive role that the self – determination of human plays in this society and how to protect the self – determination in human gene technology becomes the key point of how to insist human freedom guardian in front of the technology.

On the respect of reserving human's right to self – determination, the first point should be discussed is that if people have the right of voluntarily choosing "any" human gene technology under the situation of understanding all the risks and technology, especially the right to take germ line gene therapy or gene augmentation. The root of the problem is relating to the ownership of genes. Undoubtedly, man can be the owner of his own gene because from the purely traditional property right point of view, gene is same as cells and organs which belong to human themselves before separated from body. But this dispute is particular because gene is not exclusive to a one person but shared with family members like parents and children, even other people beyond families and all human beings further, which is different from other cells and organs. Thus, the recombination and operation of gene will not merely involve the interests of someone but of group and future generations. As the distinguishing duality of gene, the legal protection of gene not only preserves it as a independent individual possessions but reserves it as human common heritage. The ownership of gene is presenting a multiple tendency that it is no longer confined to "who I am", but extend to "who my descendants are". As a part of the whole, individuals are living as a representative of an important collective. Legislation restricts personal choice for protecting common interests. "Human are protected as a gene holder, the right of genetic identity protects the benefits of future individuals and human species over time." [1] This is the benchmark in the legislation of international world that human gene is the common heritage which belongs to all human kind and no one is allowed to make modification that may influence future generations.

But it is not enough only to adhere to the benchmark mentioned above, many scholars have cast doubts that this legislation overemphasize the point of human common heritage and integrity and diversity of species, neglecting the autonomy selection based on personal rationality. All the evidence concentrated express in three aspects: firstly, the database of human gene is unstable that the quantity and quality of the database is not unchanged with the incessant evolution of human being. Actually, integrity and diversity is not stable without calculation by a immobilized and programmed formula. Therefore, it is hard to stand up logically if only judging the legal values by keeping the integrity and diversity of human gene pool but opposing people to make autonomy choice on gene therapy. Secondly, in some circumstances, it may damage the options of present generation to attach too much importance to the options of future generations. Hence some scholars insist that the principle of autonomy of contemporary should be respected and parents enjoy the freedom right to choose whether undergo the germ line gene therapy for fear of the continuation of genetic disease. Under customary theory of capacity for civil rights, capacity for civil rights of a natural person begins with birth. And the legal definition of birth is independent from mother and capable of self - breath. [2] In existing laws, the unborn child does not have the legal right to choose but parents have. Thirdly, human have the right to enjoy better gene. Although it can not be clearly defined what is good gene, people are

---

[1] H. Boussard, "Individual Human Rights in Genetic Research: Blurring the Line between Collective and Individual Interests", Therese Murphy (ed.), *New Technologies and Human Rights*, Oxford: Oxford University Press, 2009, p. 259.

[2] H. T. Engelhardt, Jr., "Human Nature Genetically Re - Engineered: Moral Responsibilities to Future Generation", Emmanuel Agius and Salvino Busuttil (eds.), *Germ - line Intervention and our Responsibilities to Future Generations*, Netherlands: Kluwer Academic Publishers, 1998, pp. 51 - 66.

entitled the right to change their genes for specific oneself needing.[1] The choice which is completely out of self – interest is also in conformity with the principle of human dignity. Human dignity bolster up people to freely make choices out of rationality because the individual choice of demanding changes of heritable gene does not necessarily influence the integrity and diversity of all human gene but perfecting it instead. The overall ban of the practice of changing gene may cause negative critical point[2] of gene in science. In front of these questions, some international documents have made space for broader individual choices that may exist in future gene technology. In European Oviedo Convention, keeping gene integrity has not been regulated as a general right; and the interference of germ cell was not clearly banned in the World Declaration on the Human Genome and Human Rights.[3]

On the contrary, more people believe that the theory of inter – generational equity should be introduced into the discussion of human gene technology. The right of gene has the same importance as environmental right which will influence the interests of offspring and should be protected from the perspective of public reason, not just from the perspective of respecting individual choice because the latter may cause damage to all man kind including the dignity of future generation. The present generation have the moral obligation to take consideration of our offspring. And their rights of equality, options deserve protection on integrity and diversity of gene and no one can make choices instead, even though their parents. It is on no account to change

---

[1] Hermann J. Muller, *Out of the Night: A Biologist's View of the Future*, New York: Vanguard Press, 1935, pp. 113 – 114, Stuart F. Spicker, "The Unknowable Effects of Genetics Interventions on Future Generations", Emmanuel Agius and Salvino Busuttil (eds.), *Germ – line Intervention and our Responsibilities to Future Generations*, Netherland: Kluwer Academic Publishers, 1998, p. 149.

[2] Negative gene critical point derived from gene mutation, because almost gene mutation is negative and can pass on to the next generation. There are on average 8 negative genes of a person and a critical point will come to surface accumulating to the eighth generation. Based on this theory, some scholars claim that it is of great importance of improving gene quality to take on positive reproduction intervention.

[3] R – M Lozano, La Protection Européenne des Droits de l' Homme dans le Domaine de la Biomédecine, Paris: Documentation Française, 2001, p. 227.

heritable gene with relevant technology of gene therapy. Therefore, the relevant legislation should not be confined to the interests of contemporary but with a forward-looking vision. The current balance of options among generations is still in dispute, but it is with no doubt that to protect the options of present and future generations as a whole is more in line with the requirements of gene technology on the respect of guaranteeing human dignity. As the most desired and controversial technology in current genetic science, human gene technology needs to be regulated by closer legal regulations and guide, directing to the development of preserving the right to self-determination rather than violating.

Through discussing and analyzing human dignity as the legislative purpose of human gene technology, the complicated and volatile relationship between science and human nature begins to surface. Science is required to respect human nature but is not able to explain it and jurisprudence is required to set standards in a neutral position but the laws give human dignity the first place to challenge the science. Thus, the respect for human dignity not only exists under the premise that the human's cognition of human gene technology and technical quality need to be improved, but also should be upheld as a long and sustaining principle, which is not only for preserving human's need as the special subject in social life, but also a necessary condition for ensuring long social stability.

Specific embodiment of human dignity in the legal regulations of human gene technology

Human gene technology legislation is getting more and more important in the future development of society and technology. People are constantly paying attention to this problem and the depth of being influenced by the problem is extending. It is accepted that modern science and technology is playing a constructive role in providing health, wealth, and happiness for human through efforts in reducing disease and deformity and giving new opportunities for thoughts and behaviors, and it is expected to play a same role in the expansion of human nature in the future. Out of this consider-

ation, human began to seek gene therapy to deal with genetic defects. [1] People are looking forward to therapies and prevention to disease that human gene technology can bring. And the specific impact and question this longing brings is not only out of distrust of technology, but also the larger part is out of the upholding and pursuing the ethical baseline. As a way of regulating, the basic starting point of law is not only from the overall perspective, but also from the perspective of assuring human's well-being, which is one of the fundamentals of modern countries. Thus, the future of human gene technology will be bright as long as it is from the perspective of assuring human well-being and the law will do its utmost to ensure the progress and development of the technology within the range above mentioned, which is precisely the foothold of upholding human dignity as the legislative purpose of human gene technology.

The protection of human center

What exactly is the reason that people are against the application of human gene technology? Is the technology trying to make human without foresight the God, or the technology has broken the natural human form? Is the technology challenging the the basis of natural person in traditional ethics, or restoring the molecular state of human? The answer of these questions is the best interpretation of the legislative purpose of human gene technology.

Fist of all, does the opposition of human gene technology fall on technology itself? Since the humanity entered the civilized society, science and technology are playing an important and irreplaceable role. It should be realized that no all past technological revolution, no human life today. Each technological development impacted the mainstream ethics of that time, there are many scientists upholding scientific spirit but becoming a martyr such as Copernicus and Galileo. So, in today's so-

[1] Gregory Stock and John Campbell, *Engineering the Human Germline: An Exploration of the Science and Ethics of Altering the Genes We Pass to Our Children*, New York: Oxford University Press, 2000, p. 104.

ciety which is based on freedom, equality, democracy and law, there's no reason that people can not forgive the further development of a scientific civilization. Especially the human gene technology has the unique advantages to improve the rates of conquering some kinds of disease, prolong human's lives and increase the probability of human health. Besides, there exists no absolute security in the world of science and no technology is without risk. According to the law, people have to be voluntary to be involved in biomedical research because the main purpose of the research is not necessarily for promoting their health and their accession may bring immeasurable risk to themselves. Because of the company with risk, it is hard for technology to get its precise identity, like the exploration and use of nuclear power, even now, no one can give a exact conclusion whether it is good or bad. But no one will deny the value of the technology, per se, whether in economic payoff or in the complicated international competition. Therefore, the hostility of human gene technology is not onto the technology. In other words, the technology itself is without morality, rights capacity and conditions to be excepted.

Moreover, is the view against human gene technology aimed at natural state of human and maintenance of human gene diversity? Through the ages, human's attitude towards nature is wilful, supposing that all the anti – natural is anti – fact. Human has given numerous definitions of the nature, as a whole, the nature is a kind of state and regulation which dictates human's lives. From this respect, the opponents think this technology has influenced human descendants, biological diversities and human's natural states and so on. But are all the problems coming along with human gene technology? If we look at the natural state of human from the perspective of natural revolution, the problems will easily be seen. What is the natural revolution is obviously opposite to the artificial revolution. And the natural revolution is a process of human's reproduction following survival instincts without any human factor. However, it has been a long history that human controls the birth of their offspring, just from the western countries, there are far from few infanticide on the historical records from the ancient Greece and ancient Rome, even if in the Christian Middle Ages, the forbiddance of infanticide were clearly recorded in religious belief but this problem had not been fundamentally improved. Moreover, the western popularity of infanticide didn't control the quantity of babies but quality. In the military state like ancient Sparta, the requirements of infant health quality made it a certain abandonment and death to new – born babies. Besides, human is no longer in the so – called natural state since entering into the society. People getting into groups for survival entered the civil society, forming a country. They were invariably changing the natural state. Human's revolution is no longer subject to the Darwinian "survival of the fittest in natural selection" process which was mentioned in his On the Origin of Species. Human has modified some factors like food and house conditions, culture and education, leading to a front – changing of natural laws. And natural reproduction is becoming untenable in human evolution.

Therefore, people's control of self – reproduction can be detailed to every aspect and try to achieve perfection in the applicable range. Human institutions, especially the establishment of the marriage system, is a fundamental manual intervention to the human reproduction. The occurrence and popularity of some customs, like the ban of the marriage for the couple with same family name or marriage of equal status or prestige, is a kind of restriction of human multiplication, not to mention the control of child – bearing age, the treatment of sterility and the limitation of birth number and so on. It has been a long history of these control of human multiplication, with its appearance and disappearance independent from human gene technology. What's more, it is never regarded as a new subject that parents determine their children's life. Every aspect like the time of pregnancy, the gender of the baby, the growing process of the child, future occupation and life statue and so on, are all inevitably influenced and decided by the parents. Thus, human gene technology is not the main cause of parents' options to their children, but only expanding it. From this perspective, it can not be the central cause to support those anti – HGT. Even more important, in the control of human, undoubtedly, almost no country deprive human's rights of voluntary birth control and provide some technical supporting measures especially in medicine and devices. Reviewing the history of mankind, almost every moment, human beings are displaying their self – control in the process of the whole revolution. It is also because of the control, genetic determinism can germinate and grow gradually. And since the human world is theoretically divided into the natural world and the rational world. From the beginning of birth, it is a process of using rational world to control the natural world, mainly about the process of control the body by mind, even the methods developed by the rational, like plastic surgery, making up which can change the natural appearance of human beings. Once acknowledged the man's subjective initiative, it can not be denied that it requires people to adapt to natural conditions as well as to change the natural. The history of development of technology is also the history of human changing the nature, with the change inevitably effecting human itself. It was, from this aspect, one – sided against human gene technology from the perspective of following the nature.

On the other side, the natural evolution is not as what idealistic scholars argue as species diversity and better biological environment. In fact, compared with the existing time of the earth, it is not long that human occupied the dominant position in the earth. Before the time that human tried to control the nature, it already was a general trend of the extinction of plants and animals species. So far, almost 99% species ever living on the earth were extinguished. In other words, humans should not take lead responsibility for species extinction and damages of ecological environment, but what human should do is to stay the progress of deterioration. Because of the surplus oft the male giraffe, in 2014, a Danish zoo executed a healthy full – grown male giraffe in accordance with the European law of species optimization. According to the provisions of the law, the zoo should take the responsibility of assuring the quality of giraffe. When introducing a female giraffe, the number of the two ever – existing male giraffe must be reduced. Under the situation that it is difficult to find a place eligible for adoption, the zoo decided a final implementation of the execution measure to the male giraffe in order to transmit the good gene. Removing the consideration of various related real problems, for example, it has great obstacles to release the breeding giraffe to the wild because of its poor adaption to the natural environment, only from the problem itself, whether human have the right to artificially filter the species. According to the current legal provisions, the filtration of species, as a measure of assuring biological diversity, are protected by the law. Wherefore, only from a perspective of protecting biological diversity, without the regarding of the peculiar position of human, the opposite view of human gene technology can not be developed. On the contrary, many of the scholars who maintain a positive attitude of the human gene technology believe that genetic modification can make contribution to prventing the degradation of human beings, guaranteeing the whole life – sustaining of human. From this perspective, human gene technology is exactly conducive to the development and continuity of human being and to the keeping of human natural states.

Finally, the question is whether the opposing views of human gene technology is against the dominant position of human. Starting from the theory of contemporary countries and social construction, a central argument could be found in every school is that human rationality and subjectivity, except for the later, all the systems, laws and economic mechanisms can not be established. From this, the view of the anthropocentric has to be mentioned, which has lots of debate all along. Still, the foundation of current mainstream versions lie in anthropocentrism. They asserted that the anthropocentrism can be a type of cosmology, weltanschauung or ontology, that is "the anthropocentric of the universe" or "the anthropocentric of ontology" according to the anthropocentrism based on geocentricism; or "the anthropocentric of theology" according to the anthropocentrism based on teleology; or "the anthropocentric of biology" according to the anthropocentrism based on biological nature (as, organism must be self – centered); or "the anthropocentric of axiology" according to the anthropocentrism based on the nature of value (as, only humans have intrinsic values); or "the anthropocentric of epistemology" [1] according to the anthropocentrism based on epistemology (as, humans' thoughts determine the moralities created by human). All types of the anthropocentrism are centered on the the way of human to treat other species and environment in the nature including how to guarantee the hub – position of human. There has been a long negative debate about the anthropocentrism, but still not ceased, on which the ecological ethics brought the greatest impact. It requires to use "consistent link of ethics" to prove anti – anthropocentrism in order to demonstrate the consistency between human value and biological environment from the perspective of guaranteeing the ecological balance. So far, the theory has failed to fundamentally establish the human uncentred position because through the long development of western philosophy, the difference between human and animals can not be easily rooted out although today, people take it as a compulsive to protect animals even believe that animals should be given "personality rights". Scientifically, there exists some difficulties on how to prove that animals have the same rationality as human; besides, if it was proved, the current clinical trail research that stared from animals will be put into a legal freefall that human also could be the start point of clinical trail research.

From this perspective, it can be asserted that human gene technology shakes up the foundation of anthropocentrism. In human gene technology, human exists as molecules and it can be identified whether human owns real reason on a genetic level, which will make human lose the prior position to animals and may degenerate to a experimental material. And this will not only be reflected in the area of medicine and research, but also on the human social level as well. Once the central position of human is shaken, the first and foremost is the change of affirmation of human's rights,

---

[1] Wang Haiming, "Discrimination of Anthropocerntrism and Anti – Anthropocerntrism", *Journal of Liaoning University* (Philosophy and Social Sciences), Mar. 2006, p. 1.

the protection of animals and animal liability for tort in laws and with it comes the change of legal relations, like marriage, family, succession, etc, which will basically change the appearance and developing directions of human world. Although it can not be confirmed that whether the changes are kinds of progress, the current situation has no conditions to change relevant preparations. Thus, the objection to human gene technology is concerned with the central position and subjectivity of human not being shaken. From this point of view, only on the premise of protecting the subjectivity of human, the legislation of human gene technology can make contribution to the development and stability of current human society.

The stability of weighing system

It is because the current objectors of human gene technology suppose that the technology may shake the centrality and subjectivity of human. Thus, the countries worldwide have adopted sound judgment system of the value of law in the legal regulation of human gene technology by safeguarding human dignity, taking a careful and conservative attitude for protecting the centrality and subjectivity of human before the occurrence of more accurate and safer researching statistics of human gene technology.

First of all, as the legislative purpose of human gene technology, Libertarianism is major supplemented by that of Communitarianism. From the international legislative documents of human gene technology, it has always adhered to the guiding ideology of legislation centered on core values of liberalism. Many iconic words of liberalism such as individualism, freedom, egalitarian and democracy, can be searched in the documents related to human gene technology. Even in the protection of specific rights, the setting is stick to the self – centered standards. Some specific principles in the human gene technology law like informed consent, the principles of voluntariness, principle of non – discrimination, non – maleficence and beneficence all are the responses of principle of human rights required by the Libertarian Position. The article tenth of Nuremberg Code clearly defines the principles of voluntariness, informed consent and anti - damage, to which the specific provisions are assembled to ethic principle in Belmont Report, respect for persons, beneficence, justice as well as the informed consent, risks, benefit assessment and the choice of researching objects in appliance. They all have reflected the application of the basic belief of liberalism under the legislation of human gene technology. The declarations of the ethics committee of the International Human Genome Project showed specific embodiment of the adherence to the basic principles of liberalism. For example, in the Declaration about DNA Sampling, Controlling and Accessing, the ethics committee of HGP started to provide proposals on the ethical principles of the specific points of human gene technology, showing a process of the development of ethical principles from the macro to concrete. The declaration states, generally, is beneficial for the prevention and therapy of diseases to progressively research into the gene technology. However, when providing and choosing the information, the benefits of family members can not be ignored, both of timely and potential. The respect of individuals or families can be promoted and protected through trying to avoid the utilization and dissemination of identifiable samples. The protection of data is critical. The gene information created by genetic research is essential to the inheritance of immediate family whether about the legal definition of family or different social and cultural constructions. Hence, the respect for family members is of great importance. Meanwhile, the utilization and dissemination of the samples also involve the rights of the third party, such as employers, insurance providers, schools and governments in which some discrimination issues may occur. In these cases, it is indispensable to regularize the procedures of DNA sampling and accessing and to guarantee the security of samples. Out of the consideration of above problem, the ethical committee of HGP particularly expected to establish the respect for the right of freedom and informed consent as well as the choice and the respect for DNA sampling, saving and using as well as the privacy of DNA as the cornerstone of gene researching moral integrity. On one hand, it reclaims the importance of pursuing science knowledge for human progress and relief of human sufferings. On the other hand, it requires that the pursuit must accord with the standards of international human rights. It is the prerequisite of morality to respect for personal values, family need and cultural difference as well as to have the possibility of free withdraw the consent of participating the research.

However, as a brand new and of potential impact to all humankind technology, only a legislative standard is not efficient. Thus, in the legislation related to human gene technology, the requirement of communitarianism, that is "the common good". Back to the year 1995 the ethical committee of international HGP established the four cardinal principles of human gene technology in the Declaration about the Proper Conduct of Gene Research, which is ratifying the human genome as a part of our common heritage, insisting on the international standards of human rights, respecting the values, traditions, cultures and integrity of participators as well as accepting and upholding human dignity and freedom [1]. The four principles became the cornerstone for balancing the scientific and economic interests and ethics on the problems of the legislation of human gene technology. Meanwhile, the principles also fully manifested the the core human rights required by liberalism and "the common good" advocated by communitarianism as well as the high appreciation of community and integrity.

The pure individualism still has not been completely upheld in the legislation of human gene technology, with the regard of the association of gene and the interests of community, family and all mankind. It clearly stipulates in the Universal Declaration on the Human Genome and Human Rights that with HGP as the common cause of all over the world human beings. The importance of international solidarity and cooperation is self - evident. Every state plays a very important role in the cause because it takes the responsibility of promoting the solidarity and cooperation on the internation-

〔1〕 http: //www. hugo - international. org/img/statment% 20on% 20the% 20principled% 20conduct% 20of% 20genetics% 20research. pdf, the latest visit is on Feb. 19, 2014.

al stage, especially for those people susceptible to inherited diseases or already be affected as well as those families and groups, focusing on the research of the identification, prevention and therapy of inherited diseases, particularly rare diseases and local diseases which have abused large world population. On the other hand, all states around the world shall respect for the formulation principles of the declaration during the international cooperation, pushing forward the international spread of the knowledge about HGP, human diversity and genetics research and enhancing international cooperation in this area, which includes developed and developing countries. And the cooperation should guarantee the special developmental situation of the developing countries and take their development and scientific research ability into consideration to ensure that developing countries could benefit form the scientific research and apply it to the economic and social lives, which is universally benefited. Besides, the international organizations also should take a proactive role in promoting information communication among countries.

At the same time, developed as the human gene technology, the benefit sharing of HGP became a focus of public concern. It became a new project that how to make a project involving all human benefit all mankind. In the year of 2000, the ethical committee of HGP released a declaration on benefits sharing of gene, in which the basic principles and relevant pointed about benefits sharing were involved as well as corresponding suggestions. For realizing the goal, the committee has done a lot of researches on historical background, proper definition of community and the belief of human common heritage as well as the principles of solidarity and justice. The committee not only mentioned that human had witnessed a new international consensus in the past ten years, that is, researching groups should keep within the minimum and acquire some knowledge, but also strengthened that community as a term would never be separated from the attitudes of outer social world. Furthermore, on the issue of common heritage, as a species, all human share the essentially same genomes which are allowed to carry on reproduction among all human beings. From the aspect of collectivity, genomes is the common heritage of mankind and it should be upheld that the sharing benefits of human genetic heritage exist above individuals, families and groups during the application of human genetics. Therefore, the HGP must be a cause benefiting all mankind and many future prevention and therapy will be based on the genetic knowledge. It is of the best interests for everyone to work towards the best health achievable for all by powerful and rich countries as well as commercial entities. This interest in the aspect of sharing benefits is different from the interest in the sense of finance or business. Judging a interest depends on needs, values, priorities and cultural expectations. Therefore, the ethics committee gives a final recommendation: ①all mankind are equal to enjoy and acquire the benefits of gene research. ②the benefits are not limited to the individuals involved in the research. ③prior consultation with the collective or community is needed on sharing benefits. ④the immediate health interest should be provided which is decided by community need even though without any profits. ⑤all researchers should have the access to the information of general researching findings and deserve appreciation. ⑥profit entities should donate a certain percentage (such as 1% - 3%) of annual net profit to health care infrastructure and humanitarian projects. And this attitude towards human gene also was manifested in the Statements on Human Genome Project. The ethics committee of HGP put forward some proposals about the database of human gene: ①Human gene database is global public interest. a. Useful health knowledge is available to all human beings; b. Human gene database is a public resources; c. Everyone has the right to share and acquire the interest of the database. ②Individuals, families, communities, commercial entities, institutions and governments have the obligation to promote public interests. a. Public participation is the prerequisite of public duty; b. The acquirement of social interest like health care, education and decent standard of living, will boost this duty. This is the consideration of communitarianism in the legislation of human gene technology.

Subsequently, human dignity, as the legislative purpose of human gene technology, ensures the equal realization of the social level. Human gene technology makes human live in a molecular state, the gap between people ritualized and internalized, which impacts the concept of all equals from the roots. The conclusion about equality drawn from the the discussion of individual perspective must be unequal in front of human gene technology. Because from a perspective of personal gene, no one has the exactly same gene arrangement with other people, based on this inborn difference, the point that men are created equally can not be derived. Human gene technology makes the natural state that constructed by liberal scholars become the fact that can be a kind of scientific proof. Thus, we should obey the logic of funding fathers of liberalism to demonstrate equality, putting the equality on a social level. The ultimate question of equality is about the right of individual, as a member of collective, which can be enjoyed and distributed in many aspects, like social wealth, opportunities and resources. But there still exists disputes of the equality on the social level, as Dworkin mentioned: "Libertarian conceptions equality suppose that people have 'natural' rights over whatever property they have acquired in certain canonical ways and that government treats people as equals when it protects their possession and enjoyment of that property. Welfare – based conceptions, on the other hand, deny any natural right in property and insist instead that government must produce, distribute, and regulate property to achieve results defined by some specified function of the happiness of welfare of individuals." [1] The argument of whether treating equally or distributing unequally put equality to the public. The solution in human gene technology legislation to the argument is, taking non – discrimination and privacy as one of the important legislative principles to keep the equal right of individual gene on the social level. The first article of European Oviedo Code provides that: "the contracting party has the obligation to protect the dignity and features of all mankind and ensure that everyone can receive the equal respect of integrity, basic freedom

〔1〕 Ronald Dworkin, *Law's Empire* (Beijing: Encyclopedia of China Publishing House), 1996, p. 264.

and other fundamental rights in the application related to biology and medicine. In the European "Act on Protection of Individuals with regard to the processing of Personal Date" provides that every state and nation shall respect their fundamental rights and freedom, especially the right of privacy, and shall play a role of promoting the progress of economy and society, expanding trade and increasing individual benefits. Meanwhile, on the data use, it is required that free flowing and sharing among contracting states should be ensured. This way of treating equally manifested an attitude towards personal privacy and the protection of human dignity in the equality of human gene technology legislation.

The legislation of human gene technology has taken the difference among countries and communities into consideration on choosing the subjects. It clearly stipulates that equality and anti – discrimination should be realized when choosing the subjects. The International Ethical Guidelines on Biomedical Research Involving Human Subjects is a research from resource – poor groups and communities, targeting on the equal distribution of burden and interests among the subjects and discussion how to ensure the equality on the human gene technology from many perspectives, for example, the research related to vulnerable populations and children. Moreover, from the aspect of granting the human gene technology patent, the related legislation, from the angle of human common heritage, prior use of the subjects and making grant awards, hope to get benefits from the risks and ensure the equal rights of different groups when enjoying the results of human gene technology. In International Ethical Guidelines on Biomedical Research Involving Human Subjects, it also clearly regulates that the subjects are entitled to get compensation of free medical care and money if get injured during the experiments. And the dependence of the subject has the right to get compensation if died in the experiments. The compensation rights can not be abandoned as required. [1] When research involves the sponsors of the experiments, the sponsors are required to provide essential medical care for the safeties of the subjects. And it provides immediate therapy if needed to ensure the reasonable access of the results of the research to relevant communities or groups. Those measures help to realize the principle of equality.

The ethics committee of International Human Genome Organization made a more direct regulation for the equality of human gene technology by the Statement about Sharing Benefits. The statement clearly referred to the justice, and pointed out that justice is a core problem. On the problems of sharing benefits, there are at least three different meanings of the definition of justice: ①Compensatory justice : this means that individuals, groups or communities should be compensated, as the return for the contribution. ②Procedural justice: this means that, it is fair and inclusive through the distribution of compensation or decision - making process. ③Distributive justice: this means the fair allocation and acquisition of the resources and commodities. At present, on the research direction and priorities as well as the sales and enjoyment of its benefits, there is a huge inequality between rich as well as poor countries. When there is a big power gap between investigators and participants and the possibility of a substantial profit. We should support the demand for health care to respond the desirability and allocate some profit out of the consideration of justice. The ethics committee has pointed in the Statements on Human Genome Project that we should encourage the free flow of data and the benefits of fair and equal distribution of researching the database; individuals, families and communities should be protected, from discrimination and insult; researchers, institutions and commercial entities enjoy the rights to obtain a reasonable return as a way of intellectual property rights and economic contribution of database's rights.

---

[1] Han Yuehong, *Eds, Safeguard the Dignity of Life—the Research on Ethical Problems of Modern Biological Technology* (Beijing: China Renmin Press, 2005), p. 46.

Finally, human dignity served as the principle of human gene technology legislation, making the combination of physical freedom and metaphysical freedom. In the view of the conception itself of human dignity is part of will rather than those concrete ones that can be seen or felt. The view of individual autonomy or freedom is grounded on the transcendental part of the moral philosophy presented by Kant, from the perspective of free will, granting the privilege superior to any other earthly rights to freedom. The freedom mentioned above is much closer to the freedom that the classical German philosophy is always seeking for, and is of metaphysical category. In Kant's opinion, universal legal rights refer to "The summary of conditions that one person's free will shall in compliance with a universal law of freedom and keeps in line with another person's free will." [1] While Hegel once said that any existence, as well as the existence of free will, is called law. That is why law acts as rational freedom. [2] The conflation of freedom and law is the purpose of metaphysical ideology. For one reason, this pursuit causes the argument of human dignity as legislative principle of human gene technology. For another reason, it cannot be ignored in case that without the foundation of metaphysics, there will be no way to prove human dignity and freedom. Not to mention that some scholars have proved through scientific experiments and statistics that free will is just an illusion [3]. During the fight against determinism, only if the argument for freedom on the ground of philosophy is sustained, the classical position of human will be established radically, serving as

---

[1] Kant, *Immanuel Kant—Complete Works*, Vol. 6 (eds. Li Qiuling, Beijing: China Renmin University Press, 2007), p. 238.

[2] Hegel, *Philosophy of Right* (Beijing: The Commercial Press, 1979), p. 36.

[3] American psychologist Benjamin Libet, Patrick Haggard, and German neurologist John Dylan Haynes and other scientists, through the research of human brain, draw a conclusion that the brain activity can display the choice of the person between two options and it is earlier that the person's will. Haynes thought brain will be spontaneously active without any consciousness and this is a preparing process for the following decision not the process of thinking that people general believe. In another word, there is no so – called free will and all the decisions can be known by the activity of brain. Fei Duoyi, "Free will: Illusion, Disputes and Answers", *Studies in Dialect of Nature*(*Dec.* 2010), pp. 18 – 19.

the basis of legislative protection.

However, the metaphysical freedom cannot be put into practice, therefore through legislation of human gene technology, the metaphysical freedom must be transferred into real freedom. While such transformation rests on the regulations on the principle of voluntariness and informed consent of relevant legislation. On informed consent, The International Ethical Guidelines on Biomedical Research Involving Human Subjects has regulated precisely that, the secure of informed consent shall combine with supplying of basic information to the potential subject, including information about their voluntary determination to take part in and their freedom of refusing, without any detrimental influence, the end of and the risk of the research, eventually whether the subject could benefit from it and how, etc. At the same time, it explicits sponsors and researchers' liability on securing the right of informed consent, involving no unreasonable fraudulence, unjust affect or intimidation; before securing informed consent, full information should be provided and be completely understood, and the letter of consent shall be signed by every subject, and ethics committee's approval shall be given if there are any exceptions; if there are any changes in the research, the letter of informed consent of every subject shall be changed; and during long – term research program, renewal of letter of consent of every subject shall be signed in the designed internals, regardless of the research has changed or not. Those regulations of completion and compensation for informed consent, fully showing respect for human, with extreme precise manipulation, are readily accessible in practice. The Interim Measures for the Administration of Human Genetic Resources of our nation also regulated some rules over informed consent, "in the process of application and endorsement, informed consent has been stipulated explicitly, when processing the approval of international cooperation program involving our national human genetic resources", "the proof materials on informed consent of the supplier of human genetic resources and their relatives" shall be attached. Application is not permitted when there is no informed consent. Those regulations are all the reflections of putting metaphysical freedom into practice, and the measures taken to protect real freedom under the guide of human dignity, finally achieving positive effect.

Human gene technology must be objective as a scientific technology, even though its development and achievement are filled with imaginations. However, only if all those imaginations come true, it could become real science. Once human gene technology is put into social life, it will suffer all the tests by ethics and laws, and as a technology that get human involved, respect for human itself must be involved as well and influenced by it. It is not only because that the present political, economic and legal system is built on the basis of reasonable human, and mostly because that the sense of human cannot accept increasing risks out of their control. In the present legal system, cautions and restrains that human take to human gene technology result from sticking to human dignity. On one hand, it is the specialty that human dignity is grounded on human, and it is the basis for establishing social system to keep human at the dominant role; and on the other hand, legal regulations on this field still serve the end of protect the benefit of human kind. It is the point that human dignity stresses on. Therefore, human dignity could become the principal of relevant regulations about human gene technology, though it is disputable, it is also indispensable.

THE ASCRIPTION OF RIGHTS TO FROZEN EMBRYOS AND RULES ON EXERCISE OF RIGHTS

LI Yan, JIN Genlin[1]

Abstract:

Frozen embryos are created from the process of applying the assisted reproductive technology of in vitro fertilization and embryo transfer. Relevant disputes on the ascription of rights to frozen embryos and exercise of rights arise increasingly, which are much in need of being resolved. The three existing theories of frozen embryos include the subject theory, the object theory and the mediation theory. The frozen embryo possesses both the attribute of the object of real right as a special thing and the attribute of the object of personal right as personality interests determined by reproductive autonomy, which contributes to the duality of its legal attribute. The husband and wife intending to be a pair of parents by receiving the surgery of creating the embryo are rights holders, enjoying the limited ownership and rights of self – determination to the frozen embryo, which should be exercised by consensus and restricted to laws as well as public order and good customs. As successors enjoy limited rights to inheritance, personality interests and limited rights of disposal to remaining frozen embryos after the death of the husband and wife, the remaining frozen embryo adoption system after the death of husband and wife would be suggested to protect parties' interests.

---

[1] Li Yan, Doctor of Laws, Professor of Shandong University of Political Science and Law. Jin Genlin, student of Shandong University of Political Science and Law. Huang Dingquan, *Medial Law and Bioethics*, Law Press, 2007, p. 408.

I. Statement of the problem

"Under normal circumstances, humans experience the magical natural procreative process of ovulation, coition, conception, uterine pregnancy, and delivery". However, since many couples cannot procreate naturally due to multiple factors, the emergence of the assisted reproductive technology that includes the in vitro fertilization and embryo transfer technology, enables the infertile couples to have an opportunity to give birth to a child. In vitro fertilization works through taking an egg that would be put into an incubator later out of the female's ovary by professional means, and then removing the sperm from the male, separating the semen and sperm, finally putting the culture dish where the semen and the egg are placed into the incubator. "The meeting and combination process of the egg and the sperm is usually called fertilization". [1] In this process, "the egg taken out of a mother's body is fertilized by the sperm and then changes into a fertilized egg or a zygote, after that, it splits into a quasi embryo, namely, the less than 14 days old embryo. The quasi embryos are placed in a mother's uterus when they develop to the stage of 2 – 8 cells. In vitro fertilization would succeed if the quasi embryo becomes attached to the uterine lining, otherwise it fails." [2] Neonates born by means of such technology is what people usually called the "test – tube baby".

Embryo freezing technology derives from the application process of in vitro fertilization and embryo transfer technology. As The Assisted Reproductive Regulation 2003 issued by the Ministry of health provides, "no more than three embryos should be transferred in each treatment cycle, and no more than two embryos should be transferred in the first treatment cycle to women under 35". Nevertheless, success rates of embryo transfers are low, therefore, extra embryos are preserved by means of

---

[1] Feng Jianmei, "The Study on the legal issues of reproductive technology", Liang Huixing ed., *Civil and Commercial Law Review*, Vol. 8, Legal Press, 1997, p. 110.

[2] Xu Guodong, "The Study on the Legal Status of in Vitro Fertilization Embryo", *Law and Social Development*, Vol. 5, 2005.

"freezing" technology for retransfer in case of failure. Embryo freezing refers to an approach to placing the embryo and the refrigerating fluid in a cryovial to make the embryo static and preserved in the −196℃ liquid nitrogen by two cooling means of slow speed (the 2nd − 3rd day embryo) and high − speed (the 5th − 6th day blastula). The embryo preserved in the liquid nitrogen loses its activity and bears the same nature after being thawed as it does when it is cultivated. The benefit is that the fertilized egg cell can be preserved and then transferred to a mother during the natural rather than artificial menstrual cycle to increase opportunities of pregnancy. [1] Most of the infertile couples are willing to freeze extra embryos remained after the first transfer for future use. With changes of the family planning policy, young couples who are not infertile start to choose to freeze the embryos in case that they intend to have a baby in the future.

As in vitro fertilization and embryo transfer technology are being applied increasingly, legal issues of the frozen embryo start to arise, e. g. , how to deal with the situation where the infertile couple are divorced or hold distinct opinions on whether to transfer, so that they could not reach an agreement on the disposition of frozen embryos? Whether a surviving spouse may implement the transfer or not after the death of another spouse? Will successors be qualified to inherit the frozen embryo after the couple's death? Such disputes have also appeared gradually in the judicial practice of China and have caused problems worthy of study. This paper discusses legal attributes of, the ascription of rights to and disposal rules of frozen embryos in expectation of facilitating improvement of relevant legislations and judicature.

II. Legal attributes of frozen embryos

The primary problem of solving legal issues arising from the frozen embryo is their legal attributes, [2] which entail numerous domains including ethics, religions,

---

[1] Huang Dingquan, *Medial Law and Bioethics*, Law Press, 2007, p. 415.

[2] Xu Guodong, "The Study on the Legal Status of in Vitro Fertilization Embryo", *Law and Social Development*, Vol. 5, 2005.

moral feelings, population policies and laws. Professor Robert Baker of Union Graduate College devised a thought experiment to test the attitude of a person to the problem. Suppose there is a cat, an infant and thousands of frozen embryos prepared for transfer in a storage which was about to consumed by fire caused by lightning, which one will you rescue if only one of the cat, the infant and frozen embryos may be saved as a sole person standing at the door of the storage? And why? Which one of the cat and frozen embryos will you rescue if you are able to save the infant and one of another two, and why?[1] Whether the answers to the questions reflect your perception of the frozen embryo?

A. Subject, object or mediation?

The frozen embryo constitutes a legal dilemma resulting from high technology, so the controversy of it is fierce. In a nutshell, there exist three theories of frozen embryos' attributes.

1. The subject theory

Subject theories argue that the frozen embryo possesses a subject status of human. Among these theories, there is one considering the frozen embryo as a natural person or a limited natural person. It confirms that a fetus possesses the subject status of human in the first place, and there are two types of standards in identifying the fetus: the attachment theory holds that the frozen embryo has not become a human before its attachment to the uterine lining; the fertilization theory contends that the embryo turns into a person since fertilization. While the frozen embryo has been fertilized in vitro, it constitutes a subject. In the first instance of a divorce case of Davis v. Davis,[2] the plaintiff (the wife) claimed the control over the seventh frozen embryo created within the marriage so as to place it in her afterwards. The defendant (the husband) claimed to continue the embryo's frozen state till he determined

---

[1] Barry R. Furrow, Thomas L. Greaney, Sandra H. Johnson, Timothy Stoltzfus Jost, Robert L. Schwartz, *Health Law*, West Publishing Company, 2014, p. 1395.

[2] 842 S. W. 2d 588.

whether to be an unmarried father or not. The judge of court of first instance decided that the embryo turned into a "person" since fertilization and it was a baby that was actually preserved in the test tube, so that the wife enjoyed the custody and was permitted to transfer the embryo for opportunities to raise it. The first trial culminated in adoption of the fertilization theory by the court.

Such a theory seems to be obviously flawed. To start with, the frozen embryo lacks the independent personality in essence. What is more, it fails to reflect a person's intrinsic self-consciousness, which does not accord with the essential characteristic of life. "humans are entities with self-consciousness". [1] Then, the success rates stay comparatively low even the frozen embryo has been placed in a mother, which differs from life in the traditional sense after all. The frozen embryo represents just a possibility to conceive a life and is a kind of potential rather than real life—they are different in essence. The fertilization theory is rather uncommon even in the United States, which can be proved by the fact that the original judgment of Davis v. Davis has been reversed by Appeals Court. In addition, the famous American case of Roe vs. Wade had denied the statement that life begins from the moment of fertilization. [2] Finally, seeing from another angle, acknowledging its subject status will cause such a dilemma: the couple who freeze embryos will commit a crime if they relinquish or discard the frozen embryos in order to terminate the conception. Institutions where frozen embryos are stored would not dare to destroy extra frozen embryos claimed by no one, and the amount of persons in the legal sense in China would be tremendously large. Besides, such a theory would do little benefit to the development of in vitro fertilization technology. Apparently, it is inappropriate to define the frozen embryo as a subject.

"The legal person theory" constitutes another subject theory. Art. 123 of the

---

[1] Qiu Renzong, *Bioethics*, Shanghai People's Publishing House, 1987, p. 88.

[2] Roe v. Wade, 410 U. S. 113 (1973).

Louisiana Civil Code sets forth the capacity of embryos fertilized in vitro: the embryo fertilized in vitro enjoys the status of juridical person before being placed in the uterus, while the embryo enjoys the status of natural person as a fetus after being transferred. Its dominant idea is treating the frozen embryo as a person's "stale of fiction" where the embryo changes from a thing to a person, rather than endowing them with complete rights of a natural person; the embryo is solely granted the right to life, which gives rise to the right not to be hurt. Only under one condition that is provided by the law may fertilized embryos be disposed of, namely, those frozen fertilized embryos fail to develop successfully within 36 hours since they are thawed. [1]

The starting point and guiding idea of the legal person theory are worth learning, after all the embryos are seeds of life rather than things in general. The legal person theory aims at improving the legal status and strengthening the protection of frozen embryos to prevent them from being infringed by others unlawfully. Nevertheless, the legal person here is not the one usually known in the continent law system, "the fertilized embryo is a union of men's sperms and women's eggs, which amounts to the union consisted of natural persons in the name of an association". [2] The original "conception of legal person would be completely changed" [3] if such legal person theory is adopted. Embryos are not persons, let alone legal persons in the traditional sense.

2. The object theory

An object refers to "a subject matter that is essential to the establishment of rights that take visible or invisible social interests as their contents or ends". [4]

---

[1] Erik W. Johnson, "Frozen Embryos: Determining Disposition Through Contract", *Rutgers Law Review*, 2003, (5).

[2] John Bologna Krentel, The Louisiana "Human Embryo", Statute Revisited: Reasonable Recognition and Protection for the in Vitro Fertilized Ovum, Loyola Law Review, pp. 240 – 246.

[3] Xu Guodong, "The Study on the Legal Status of in Vitro Fertilization Embryo", *Law and Social Development*, Vol. 5, 2005.

[4] Shi Shangkuan, *On General Provisions of Civil Law*, China University of Political Science and Law, 2000, p. 221.

Rights are legal powers which are protected by law and may possess specific interests. [1] Such specific interests in themselves are objects of rights. [2] There are numerous forms of objects, mainly including things, the specific person's acts or legal interests, rights holder's own personality interests, products of spirits and rights, etc. [3] The property theory, which defines the embryo as property and acknowledges that it may be the subject matter of a contract or be inherited, is the main theory that considers frozen embryos as objects. In York v. Jones, Court of Virginia regarded frozen embryos as property that is a subject matter of a contract of deposit. [4] In Hecht v. Kane, the defunct had frozen his sperms and designated his wife to use them, Appeals Court of California considered sperms disposed of through a will as the "property" that may be dealt with at the discretion of the defunct. [5] In the aforementioned Davis v. Davis, Appeals Court reversed the judgment that regarded the embryo as a child of first trial, acknowledging the man's constitutional right to decide not to be a father and considering the frozen embryo as a special form of jointly possessed property, to which both parties enjoyed the right of disposal equally.

The property theory treats embryos—seeds of life—as property, depreciating the value of life. Section 90a of German Civil Code prescribes that "animals are not things" to express respect for animals, therefore, there is no doubt that treating embryos as property is retrogressive. "It is inappropriate to regard embryos as things." [6] What is more, the property theory ignores the fact that the value of frozen embryos

---

[1] Shi Qiyang, *On General Provisions of Civil Law*, China Legal System Publishing House, 2010, p. 26.

[2] Liang Huixing, *On General Provisions of Civil Law*, Law Press, 2011, p. 149.

[3] Shi Qiyang, *On General Provisions of Civil Law*, China Legal System Publishing House, 2010, pp. 176 - 177.

[4] Sina A. Muscati, "Defining a New Ethical Standard for Human inVitro Embryos in the Context of StemCell Research", *Duke Law and Technology Review*, 2002.

[5] Mason & McCall Smith, *Law and Medical Ethics*, Reed Elsevier (UK) Ltd, 1999, p. 486.

[6] Kitagawa Shantaro, "On the Legal Model in the Most Recent Future", Liang Huixing, *Civil and Commercial Law Review*, Vol. 6, Law Press, 1997, p. 294.

lies in possibilities of providing life, which cannot be compared with property. And problems like how to partition the proportion of embryos which is commonly enjoyed by couples still exist.

3. The mediation theory

The mediation theory contends that embryos are neither subjects nor objects. They represent a specially protected status rather than persons or things. Such a special status "comes from the capacity of developing into a neonate". [1] In aforementioned *Davis v. Davis*, [2] the Supreme Court of Tennessee in the third instance defined embryos as transitional things, which means they are neither persons nor things, since they are respected for "their possibilities of developing into human life". It seems that on the surface the mediation theory avoids limitations of the subject theory and the object theory, however, "establishing the model of person——mediation——thing" [3] and such "special status of capacity of developing into a neonate" which represents neither subjects nor objects are difficult to be defined within elements of legal relationships and to be placed in legal relationships to specify contents of relevant rights and obligations.

B. Frozen embryos possess double attributes of objects of real rights and objects of rights of personality

The author holds the view that frozen embryos possess double attributes of objects of real rights and objects of rights of personality. As objects of real rights, although frozen embryos are objectively existing substances, they are special things that possess "potential personality", which restricts their purposes of disposition and excludes the application of provisions in respect of property. As objects of rights of personality, frozen embryos are personality interests concerning decisions of infertile

[1] Jeremy L. Fetty, "A 'Fertile' Question: Are Contracts Regarding the Disposition of Frozen Preembryos Worth the Paper upon Which They are Written?", *L. Rev. M. S. U. - D. C. L.*, 2001.

[2] Junior L. Davis v. Mary S. Davis, 842S. W. 2d588, 1992.

[3] Xu Guodong, "The Study on the Legal Status of in Vitro Fertilization Embryo", *Law and Social Development*, Vol. 5, 2005.

couples who intend to be parents on whether to realize their reproductive rights through assisted reproduction.

1. Frozen embryos' attribute of objects of real rights: a special thing

"A thing means, apart from human body, res corporales and natural power which could independently be controlled by manpower and satisfy living needs of human society."[1] In the first place, frozen embryos are created through technical treatment of sperms and eggs taken out of a human body. Sperms and eggs have been "parts of a human body till they separate from it, since when they would constitute no part of a body but things in the legal sense".[2] As Art. 99 of The Draft Civil Code compiled by professor Liang Huixing provides, "organs, blood, marrow, tissues, sperms and eggs etc. may constitute objects of civil rights on condition that public order and good customs are not violated."[3] The provision "takes the development of science and technology into consideration, especially the development of the organ transplantation technology and the artificial reproductive technology of medical science."[4] The Draft Civil Code compiled by professor Wang Liming elaborates that a human body is not a thing, but some parts separated from a human body, such as organs, milk, blood, sperms and eggs, etc., may become things and objects of real rights. Disposition of certain parts of human body is effective, provided that it does not violate public order and good customs.[5]

In the second place, frozen embryos are products of high technology, which are stored by professional institutions. "Something that can be disposed of by means of

---

[1] Wang Zejian, *On the General Provisions of Civil Law*, China University of Political Science and Law Press, 2001, p. 208.

[2] Wang Liming, *Study on the Personality Law*, Renmin University of China Press, 2012, p. 316.

[3] Liang Huixing, *The Draft of Chinese Civil Code and Its Reasons: General Provisions*, Law Press, 2013, p. 191.

[4] Liang Huixing, *The Draft of Chinese Civil Code and Its Reasons: General Provisions*, Law Press, 2013, p. 191.

[5] Wang Liming, *The Proposed Draft of Chinese Civil Code from Scholars and Its Legislative Reasons*, Law Press, 2005, p. 242.

science and technology possesses a nature of disposal", [1] which reflects that the scope of the thing extends constantly as the development of social society and the improvement of people's abilities of disposal. [2] In addition, embryo freezing is the measure taken for in vitro fertilization, whose purpose is to place in a mother of an embryo to conceive a life. Frozen embryos have certain value or utility value and are able to satisfy living needs of human society. In the end, the frozen embryo is kind of tangible object, as a geneticist states, "the so called embryo is no more than such a thing as moss and fluff …and has no difference from eggs of any other mammals." [3] Nevertheless, it has to be noticed that frozen embryos are fertilized embryos which bear the potential possibility of developing into life and special things that contain the future life. Out of the respect for life values, such a thing differs from the general thing we usually understand and turns into a fetus when it is placed in a mother (attaching to the uterine lining), which makes it a special thing

The particularity is that ownership of frozen embryos is jointly enjoyed by the infertile couple who intend to be parents and receive surgery of in vitro fertilization and embryo transfer. Frozen embryos are not allowed to be partitioned or sold. Basically they are res extra commercium, and they are protected and restricted. The right of disposal is limited to decisions on whether or not frozen embryos to be transferred, discarded, destroyed or donated. Donating frozen embryos is feasible in theory. Since the purpose of the assisted reproductive technology is to assist infertile couples in realizing reproductive rights, legislations of many countries permit donating sperms and eggs, acknowledge the donation as a humanitarian act, and allow using donated sperms or (and) eggs from the beginning when applying in vitro fertilization and embryo transfer technology. However, legislations differ on whether frozen embryos

---

[1] Shi Qiyang, *On General Provisions of Civil Law*, China Legal System Publishing House, 2010, p. 176.

[2] Ma Junju, Yu Yanman, *The Original Theory of Civil Law*, Law Press, 2010, p. 69.

[3] Huang Dingquan, *Medial Law and Bioethics*, Law Press, 2007, p. 441.

which have been created already (for instance, extra frozen embryos remained by infertile couples who have had a successful transfer, embryos remained by couples who decide not to transfer any longer or frozen embryos left by infertile couples who are dead, etc.) could be donated to other infertile couples for direct transfer. China's current regulations on the assisted reproductive technology are The Assisted Reproductive Technology Regulation 2003 issued by Ministry of health, Basic Standards and Technical Specifications of Human Sperm Bank, and Ethical Principles of Assisted Reproductive Technology and Human Sperm Bank, according to which, any form of sales of gametes, zygotes and embryos are prohibited. Ethical principles and technology specifications are provided for donations of sperms and eggs which are acknowledged as humanitarian acts. In terms of the donation of embryos, Art. 3 of The Assisted Reproductive Technology Regulation stipulates that "the frozen embryo is prohibited from being given away", which refers to technical personnel rather than infertile couples that are not allowed to give away embryos. Nevertheless, in practice even someone would donate frozen embryos to other infertile couples and it is not prohibited by law. Embryo donation still remains unfeasible as a consequence of lacking supports of technical regulations.

2. Frozen embryos' attribute of objects of rights of personality: personality interests determined by reproductive autonomy.

The right of self-determination means "one has the right to decide on matters irrelevant to the others"[1], namely, "making decisions about one's private affairs"[2]. Self-determination constitutes the main content of rights of personality, and together with the freedom of will, it can be included in other personality interests.[3] Since embryos are developed and fertilized in vitro, the interests of women taking control of

[1] Yamada Zhuoki, "Personal Affairs and Self-determination", *Japanese Review*, 1987, p. 3.

[2] Liu Shiguo, "The Right of Privacy of Patients: the Right of Self-determination and the Right to Control Personal Information", *Social Science*, Vol. 6, 2011.

[3] Wang Zejian, *Tort Law*, Peking University, 2009, pp. 135-136.

their bodies no longer exist and couples are left only with the "right of self – determining whether to become a parent"[1], as the case of Roe v. Wade has confirmed that the freedom of abortion is within the scope of self – determination.[2] Similarly, frozen embryos are carriers of a couple's self – determination, namely, autonomously determining to procreate or not to procreate. Self – determination is "a content of the freedom of personality, supplementing the specific rights of personality and protecting new types of personality interests"[3]. Such an opinion is also reflected in the German case of the destruction of frozen sperms[4], an argument of whose judgment referred to that "the object which section 823 (1) of German Civil Code protects is the existence of personality and its realm of self – determination rather than a substance"[5]. "The separation of certain parts from a body and the reunion of the two, which is based on the will of the rights holder, aims at protecting or realizing the body's functions and is the right of self – determination of a rights holder."[6]

The personality interests of self – determination are protected by law, which is reflected in legislations of our country, such as the patient's right of self – determination prescribed by *Tort Liability Law* (Has the failure), the right of marriage prescribed by *Marriage Law*, as well as the "consent form" of relevant technology and "multifetal pregnancy reduction agreement" signed between institutions implementing in vitro fertilization and embryo transfer technology as well as its derivative technology and infertile couples. Self – determination is used to be personality interests which are included in the freedom of personality of the general right of personality, and "exclusively belongs to and is enjoyed by rights holders themselves, and shall not be

[1] Huang Dingquan, *Medial Law and Bioethics*, Law Press, 2007, p. 436.

[2] Roe v. Wade, 410 U. S. 113 (1973).

[3] Wang Liming, *Study on the Personality Law*, Renmin University of China Press, 2012, p. 167.

[4] BGHZ124, 52.

[5] Wang Zejian, *Tort Law*, Peking University, 2009, p. 104.

[6] Wang Zejian, *The Law of Personality Right: The Interpretation of Law, Comparative Law and Case Study*, Peking University Press, 2003, p. 103.

assigned nor inherited" [1], before the right of self – determination is specifically provided by law. Therefore, frozen embryos bear the attribute of objects of rights of personality as personality interests of reproductive autonomy. In the aforementioned case of Davis v. Davis, the Supreme Court of Tennessee held that either The Constitution of The United States or The Constitution of the State of Tennessee protects personal rights of privacy which contain reproductive autonomy, while the reproductive autonomy comprises of two rights of the same meaning—rights to procreate and rights not to procreate, both of which are protected and restricted. [2]

III. Rights of couples who intend to be parents and create frozen embryos by surgery and rules on exercise of rights

Couples who intend to be parents and create frozen embryos by surgery should be recognized as rights holders of frozen embryos and legal parents of children who are born as a result of the successful embryo transfer to the mother. For frozen embryos may be created by sperms and eggs of couples themselves or by donated sperms or (and) eggs, rights holders are not necessarily genetic parents of children that are born to them. Nonetheless, since embryos' and children's coming into existence are based on determinations of infertile couples, couples who exercise their rights of self – determination on creating frozen embryos are chosen to be rights holders by law.

A. The ascription of rights and rules on exercise of rights during the period where a contract of marriage exists

Based on above stated attributes of frozen embryos, when couples are still living, frozen embryos are special things that carry personality interests. In this situation, the agreed disposition of frozen embryos should be respected from the perspective of either joint ownership or reproductive autonomy, if such disposition is conduc-

---

[1] Wang Zejian, *The Law of Personality Right: The Interpretation of Law, Comparative Law and Case Study*, Peking University Press, 2003, p. 48.

[2] Barry R. Furrow, Thomas L. Greaney, Sandra H. Johnson, Timothy Stoltzfus Jost, Robert L. Schwartz, *Health Law*, West Publishing Company, 2014, 1387.

ted by consensus of the couple and consistent with the law. Since the whole process from the extraction of gametes to the development and then the placement in a mother of a frozen embryo is a reflection of consensus of both parties' right of self - determination, embryos cannot be disposed of compulsorily if any disagreement exists in any stage of the process. The Human Fertilization and Embryology Act 1990 of the United Kingdom sets forth that consensus of both parties is needed in any stage of the process of artificial fertilization. Moreover, it is provided that the first storage period in respect of fertilized eggs and embryos are such peirods not exceeding five years and which could be extended to ten years if the donator gives his consent in writing. If the couple are divorced or the donator is unable to be contacted with, which renders the written consent unobtainable, the embryo's life would be terminated. [1] The storage period prescribed by law avoids the problem of long - term storage as a result of two parties' failing to reach an agreement.

B. The ascription of rights and rules on exercise of rights when the couple are divorced

If a couple agrees with the disposition of frozen embryos when they are getting a divorce, the agreement should be observed. The consensus of placing in a woman of an embryo represents intrinsically rights of self - determination of the two parties and is supported by jurisprudence. Moreover, the placement is not so different from natural procreation, except the time of pregnancy is delayed due to various reasons, and both babies resulting form frozen embryos which are transferred after the divorce and children conceived before the divorce are divorcing parties' children. Of course, there are scholars assisting that it is kind of cruel to allow a child to stay in a single parent family once he (she) is born from the perspective of the best interests of the child. Countries and regions that are more advanced in such a filed hold conservative attitudes towards it, for instance, The Relevant Artificial Reproduction Act adopted

[1] Mason & McCall Smith, *Law and Medical Ethics*, Reed Elsevier (UK) Ltd, 1999, p. 487.

in March 5th, 2007 in Taiwan recognizes the husband and wife as subjects of artificial reproduction; German Embryo Protection Act 1990 sets forth that the artificial reproduction needs to be implemented within the marriage. However, the present law of Mainland China makes no provision for this. Considering the protection of the unborn child and the fact that China's family law and Law of Succession are based on the traditional parent - child relationships, supporting the transfer to the woman of the frozen embryo after the divorce will give rise to many new types of disputes. As a result of the deficiency of legislative and judicial experience in China, the conservative attitude is suggested.

1. Either party's interests should be considered when the divorcing couple exercise the rights to frozen embryos.

If a couple disagree with each other on whether to transfer embryos to realize reproductive rights, usually the court would not compel transfer. Balancing rights to procreate agaist rights not to procreate, the party who intends not to procreate will be supported, since those two rights are equal. Some courts will take account of factors like the more effort that women put in the process of surgery, or subsequent fertility of a party, etc. When different rights are involved, the court will be more likely to balance each party's interests. When the above case of Davis v. Davis proceeds to the third instance of the Supreme Court, the situation changed, where both parties remarried and the woman demanded that the embryo be donated to an infertile couple rather than transferred to her, while the man was willing to destroy rather than to donate the embryo. The court concluded that the man's right not to be a genetic father outweighed the woman's right to donate the extra embryo to a third person for bearing and raising a child, and that ownership of the embryo is held by the man who might realize his rights not to procreate through self - determination.

In J. B. v. C. C. which was heard by the Court of New Jersey, [1] the claim

---

[1] 783 A. 2d 707 (N. J. 2001).

was opposite: J. B. (wife) required that frozen pre - embryos be discarded, while M. B. (husband) requested the court to allow the transfer or donation to others of the pre - embryos. In this case, J. B. suffered a miscarriage after marrying M. B. and had difficult in conceiving a child. After seeking medical advice, J. B. learned that she had a condition that prevents her from becoming pregnant, although M. B. did not have the problem of infertility. They decided to attempt the surgery of in vitro fertilization and embryo transfer at Cooper Center for In Vitro Fertilization. Eleven fertilized eggs are created, four of which were transferred to J. B., with seven extra ones cryopreserved. J. B. became pregnant and gave birth to a daughter. Then the couple separated. J. B. filed a complaint for divorce and informed M. B. that she wished seven pre - embryos that they preserved to be discarded, but M. B. disagreed and demanded those pre - embryos to be transferred or donated. Through summary procedure applied by J. B., judgment of first instance and the Appellate Division of New Jersey concluded that M. B. was not infertile, and that he sought control of the pre - embryos merely to donate them to another, which was outweighed by the wife's (infertile) interests in avoiding procreation. The third instance of Court of New Jersey acknowledged the Appellate Division's opinion, holding that pre - embryos should be destroyed. However, since J. B. did not object the continued storage of those pre - embryos, the court decided that pre - embryos could be continuously preserved if M. B. wish to pay any fees associated with that storage. M. B. must inform the trial court that whether he will do so, otherwise, those frozen pre - embryos will be destroyed. It is perceived in this case that, even a party who intends not to procreate is infertile, her right not to procreate still outweighs the right of another party.

2. The validity of a contract signed by couples of frozen embryos

In aforementioned J. B. v. M. B., the parties and the Cooper Center signed a consent form which described the procedure of the surgery and contained language discussing the control and disposition of the pre - embryos. Nonetheless, if there is no agreement on the disposition of frozen embryos having been reached by divorcing parties, except that they should apply to the court for an order to dispose of embryos, the court holds that, when reviewing the validity of the contract, the two parties' reproductive autonomy would not be restricted even they desire to be parents when the contract is concluded, and that they could change their minds at any time they want, because contracts for forced procreation are against public policy and invalid. In A. Z. v. B. Z. which is heard by the Supreme Court of Massachusetts in 2000, [1] the court also concluded that contracts which force people to become biological parents are invalid because of the violation of public policy.

Certainly, the consent form or other contracts signed by the parties and medical institutions are valid only if they do not violate legal provisions or provide constraints on rights of self – determination, and the conclusion of a detailed and specified contract will prevent numerous disputes from arising. In Kass v. Kass which is heard by New York Court of Appeals, [2] the court ordered that frozen embryos should be donated by divorcing parties for scientific research in accordance with the contract when they could not reach an agreement. That is because the consent form signed by the couple and the hospital clearly states that if no agreement is reached on the disposition, the frozen embryos would be donated to the hospital for the scientific research of in vitro fertilization and embryo transfer technology. What is more, prior to the couple's divorce by consent, their marital settlement agreement specified that "if there is no disposition agreement being reached, the frozen embryos should be donated to the hospital for scientific research according to the consent form, and neither party will enjoy rights to them" once again. So the court held that the contract is valid.

C. The disposition of frozen embryos after the death of a spouse

Since rights of self – determination possess "personal specificity and cannot be

[1] 725 N. E. 2d 1051, 1057 – 58 (2000).

[2] 696 N. E. 2d 174 (N. Y. 1998).

assigned nor inherited" [1], frozen embryos bear solely the attribute of a thing to the dead one, while they are still personality interests of self - determination to the surviving one. When the disposition of personality interests conflict with the disposition of a thing, personality interests might be given preferential consideration. The disposition of frozen embryos shall be recognized as an embodiment of the right of self - determination of the surviving one, if the dead one has not objected to such a disposition specifically before his (her) death. Of course, if the intention of self - determination which has been manifested by the dead party before death is consistent with the surviving one's, their consensus should be complied with; otherwise, the surviving one may not implement the disposition.

IV. The disposition of frozen embryos after the death of the couple: rights of successors to frozen embryos and the construction of the embryo adoption system

A. Rights of successors to frozen embryos after the death of the couple

Personality interests of reproductive autonomy carried by frozen embryos are extinguished with the death of the couple. Although frozen embryos are pure objects of real rights at the moment, they shall not be subject matters of a contract of sale or in circulation, since they possess the particularity as a potential life. Rights of successors of the dead couple to frozen embryos bear dual attributes of the right to inheritance of a thing and personality interests.

1. Successors enjoy restricted rights to inheritance to frozen embryos

As a pure object of real rights, frozen embryos are valuable to successors of the couple after the couple's death. They are objective substances left by decedents. In a sense, they contain certain economic interests such as operating costs paid by decedents, the storage fee, etc., and are embodiments of the division of labor in society

---

[1] Wang Zejian, *The Law of Personality Right: The Interpretation of Law, Comparative Law and Case Study*, Peking University Press, 2003, p. 46.

and social value. However, such a kind of property cannot reflect its economic value through transaction as a result of the particularity of frozen embryos in themselves. Consequently, though the frozen embryo as a special thing could be inherited by successors, it is not full ownership that is inherited, since ownership of frozen embryos enjoyed by infertile couples before their death is restricted as well. The restricted ownership inherited by successors will be further restricted as a consequence of the extinguishment of rights of self – determination of decedents. Successors may only inherit rights of possession (rights of entrusted storage) and restricted disposal rights (determine to relinquish, donate by permission of law, may not be sold, may not be used for surrogacy, etc.). The special protection and the restricted disposition of frozen embryos set forth by law cause no damage to their attributes of things, and no provisions prohibit the inheritance of them.

2. Successors enjoy personality interests to frozen embryos on the basis of personal status relationships

Successors are only people who care about the fate and enjoy personal interests of frozen embryos after the death of the couple. Against the background of China's traditional fertility culture and the national policy of family planning, most people who receive the assisted reproductive surgery cannot procreate naturally and have no child, so that their successions mostly are elders of the family such as parents, etc. Relinquished frozen embryos may bear personality interests such as the grief and the hope of families who have lost their only child, when both parties are the only children of their family. Apparently, any non – successors, including institutions that perform operations and store embryos, do not possess such personality interests, and nobody will care about the frozen embryos left by any other irrelevant persons.

3. Restricted disposal rights of successors to frozen embryos

After the inheritance, ownership of frozen embryos is enjoyed by successors and may not be partitioned or disposed of illegally. If successors have reached a consensus, they may continue to store, relinquish, destroy, (theoretically) donate the embryos to others or for scientific research, or, deliver them to other institutions for storage based on the agreement of successors; otherwise, the embryos should be prevented from being partitioned and stored by institutions for purposes of public welfare.

As discussed earlier, even the disposition of frozen embryos by infertile couples must be consistent with legal provisions and is restricted by "compliance with regulations concerning population and family planning, the prohibition of the sale of embryos, etc. ", so that exercise of disposal rights to embryos by successors should be restricted by law as well after the inheritance. Apart from the above restrictions, The Assisted Reproductive Technology Regulation sets forth that medical workers of medical institutions and service centers for birth control are prohibited from implementing the surrogate technology or embryo donation. Rights to inheritance, disposal rights and personal interests enjoyed by successors to frozen embryos should not be denied even restrictions exist.

B. A conception for constructing the embryo adoption system

Embryo adoption refers to that the frozen embryo is donated and transferred to a infertile couple who are to adopt it when it develops into a life. The embryo adoption system can satisfy many parties' demands for interests in a case where the infertile couple are dead and their frozen embryos are relinquished. For decedents of the death, it meets their emotional needs for continuing the family line and avoids successors' moral guilt of destroying frozen embryos which are still seeds of life in people's minds. For the infertile couple to whom the frozen embryo is donated, it not only meets their needs of procreation, but also reduces the cost comparing with starting developing the embryo from donated sperms and eggs. Furthermore, bodily conceiving by a foster - mother facilitates the development of a harmonious parent - child relationship between them. For frozen embryos, having the opportunity to develop into a life and to experience the beauty of life is much more valuable than being destroyed as waste things, and the interests of possibly developing a baby from an embryo has been fully taken into consideration. As in McKay and Another v. Essex Area Health Authority and Another, [1] the plaintiff suffered from measles filed "a wrongful life claim", Ackner LJ decided against the plaintiff based on that allowing such a claim "would impinge on the sanctity of human life" [2], which as well constitutes a basis of deciding against the born disabled person who argues that non – existence is better than being born with a hand – cap in most countries. [3] Thus the embryo adoption system is "a most moral way to dispose of remaining embryos" and an excellent work resulting from the measurement of interests and the legal technique.

Attempt to construct the embryo adoption system has appeared in legislations and practice of the United States. Option of Adoption Act (2009) of Georgia for the first time establishes a double track adoption system in legislation and regards embryo adoption as a way to start a family, to transfer and to establish parental rights. [4] The details are as follows: ①Subjects of adoption are infertile women who are married. To avoid the controversy with respect to scientific ethics, homosexual persons and single mothers cannot be the subjects of adoption. Besides, the mental status and the habit (whether or not a drug abuser) of the applicant should be examined based on the "best interests of the child" principle. ②Sign the relevant agreement to guarantee that custodians' rights and responsibilities for frozen embryos are completely transferred to adopters. ③Donators are limited to people "aged under 40". ④In the adoption of embryos, 76 percent of clinics adopt a mutual blind and confidential approach to complete adoption, namely, any control taken by donators of a-

---

[1] McKay and Another v. Essex Area Health Authority and Another, [1982] Q. B. 1166.

[2] Robin Mackenzie, "From Sanctity to Screening: Genetic Disabilities, Risk and Rhetorical Strategies in Wrongful Birth and Wrongful Conception Cases", *Feminist Legal Study* 7, 1999, p. 180.

[3] Zhang Xuejun, "The Legal Application of the Wrongful Life Action", *Legal Research*, Vol. 4, 2005.

[4] Wu Wenzhen, "The Practice and Legislation of Embryo Adoption in the United States and Its Enlightenment to China", *Social Science*, Vol. 6, 2011.

dopters and children who are to be born would not be allowed. A contract signed by relevant parties involves almost all situations where future disputes may take place to nip troubles in the bud.

Despite that embryo donation is not prohibited currently in China, technical personal are "prohibited from performing embryo donation surgery", and no technical regulation is enacted. Based on the present situation of China, it is suggested that a system applied only to dead couples' embryos which may be donated by successors to infertile couples for adoption be established in Adopting Law. To avoid unnecessary controversy concerning scientific ethics, a conservative attitude is held towards the applicable objects of such a system. ①Applicants must be infertile women who are married, and embryos should be adopted by husband and wife in concert. Unmarried women, (at the time of application) divorced women, or homosexual persons cannot apply for adoption. ②To accord with the "best interests of the child" principle, the adopter's physical condition and life style should be taken into account. ③Determine the age for adoption. Adopters should be in the age bracket which is right for pregnancy from the perspective of medical science, for they have to go through embryo transfer, pregnancy and delivery. ④Apply the mutual blind principle. Information on embryos, persons placing out the embryo for adoption, and adopters should be kept totally confidential (except that both parties agree to let each other know). ⑤A contract shall be entered into for the legal transfer of rights and responsibilities for an embryo and for any child that may result from the embryo transfer, and an agreement on future disputes resolution needs to be reached prior to embryo transfer. ⑥Given that embryo adoption is an act of identity which relates to public welfare, it is suggested that relevant state organs apply the review and registration system.

In summary, the development of high technology brings good news to infertile couples, and the success rates of in vitro fertilization have increased to a great extent due to the development of embryo freezing technology. Nevertheless, such development will inevitably give rise to a great number of disputes over frozen embryos. "We can foresee that the court will teem with similar cases". [1] Therefore, it is necessary to improve the legislations on relevant legal issues of frozen embryos as soon as possible and to make laws of assisted reproduction as well as embryo protection, etc. When encountering legal gaps and legal loopholes, the court shall fill the gap by means of according with legal principles, referring to ethical norms and interpreting law in judicial practice. The consent form has to be signed between parties attempting embryo freezing technology and medical institutions to nip troubles in the bud.

[1] Huang Dingquan, *Medial Law and Bioethics*, Law Press, 2007, p. 438.

HEALTH AS AN EMBODIMENT ACROSS LIFE COURSE: FROM CELL TO INTERVENTION

Cyrille Delpierre, Thierry Lang

Abstract:

In social epidemiology, a major difficulty is to understand the way social determinants are linked from childhood to adulthood to influence health. Identifying these causality chains implies to understand how the environment, in the broadest sense, becomes biological, how it is incorporated or embodied and can modify the normal functioning of the body to favour the development of pathologies over time. In this article we will concentrate on one aspect of the links between early life and health, highly amenable to public health interventions, developed and known as social epigenetics. This approach, sounds a priority as a clash between two terms, evoking on one hand, a new development of an ever more efficient technical and biological medicine and on the other hand, social concerns reminding social medicine of the nineteen century. However, epigenetics, as one part of the lifecourse approach in epidemiology, may be an incentive to carry out new types of multifactorial interventions. It emphasizes the need to rethink the role of the early environment in the production of the health status of a population and its distribution. Investing in a favorable environment from the early years seems therefore fundamental for the health and well - being throughout life. Our ability or the ability of human beings to adapt to their environment also means that interventions are possible throughout life to improve health and well - being.

Introduction

The existence of a socioeconomic gradient of morbidity and mortality is a well - known phenomenon in Europe as in the important variability existing across countries, particularly for mortality. Studies on social inequalities in health (SIH) constitute an increasing and dynamic field of research in Europe since the 1980s, even if France is still behind with regard to the other countries such as the United Kingdom or Canada. Nevertheless, the reduction of SIH represents a public health priority in France. The current period is in fact characterized by the collective awareness of the existence of strong health inequalities in France and by an increasingly explicit request for knowledge, for developing interventions likely to reduce them. To illustrate this, we can refer to the report on SIH by the French High Council of Public Health, the appearance of the reduction of SIH as one of the priorities in the plan Cancer 2 and for the Regional Health Agencies.

The international literature regarding interventions to reduce SIH is however relatively poor and the French literature, almost non – existent. One explanation may be the difficulty in evaluating the impact of such interventions, but also the lack of knowledge on mechanisms, processes by which social determinants influence health, although the role of numerous social determinants has been shown. These determinants may have a quite particular influence on the future health status in case of early exposures, during preconception, foetal period or childhood. Adult diseases therefore may partly find their origins in early exposures to a number of noxious social determinants. These scientific advances are highlighted in the Marmot review report published by the WHO in 2008[1], which stipulates that for reducing SIH it is urgent to: "Take a stand in favour of a global approach on the first periods of life and apply this approach by learning on the existing programs and by widening the interventions planned during the first years of life to the social, emotional and cognitive development/ learning of the language".

A major difficulty is to understand the way these determinants are linked from

[1] CSDH Closing the gap in a generation: health equity through action on the social determinants of health. Final Report of the Commission on Social Determinants of Health. Geneva, World Health Organization, 2008.

childhood to adulthood to influence health. Identifying these causality chains implies understanding how the environment, in the broadest sense, becomes biological, how it is incorporated or embodied and can modify the normal functioning of the body to favour the development of pathologies over time. Explaining the genesis of SIH is not only a question of physical, chemical exposures and/or health behaviours socially distributed in the population, but also a question of how some exposures, socially distributed are likely to "directly" modify physiological processes involved in the development of diseases many years later in some groups of people, explaining a part of social gradient of health.

The identification of such mechanisms would enable the identification of potential targets for leading individual and collective interventions aiming at reducing SIH. It could also highlight periods along the lifecourse during which some interventions would be more likely to be relevant. To do such analyses, the conceptual framework of lifecourse epidemiology[1], and the specific mechanisms defined therein, namely critical periods, accumulation of risk and causal trajectories, with a particular interest in exposures occurring in the earliest phases of life (in utero, infancy and childhood) and their impact on adult health, may be relevant. Lifecourse epidemiology is an interdisciplinary approach stemming from the social sciences including biological, behavioural and psychosocial mechanisms which may all influence the development of a chronic health condition during an individual's life course.

According to this approach susceptibility to illness or poor health is the result of complex interactions between biological and social phenomena. An individual's health status is therefore the result of that individual's adaptation to their environment, which is dependent upon their own characteristics (biological, social, psychological), as well as those of their environment, which may, in turn, be influenced

[1] Kuh D, Ben - Shlomo Y., *A Life Course Approach to Life Course Epidemiology*, Oxford: Oxford Medical Publications, 1997.

by factors such as socioeconomic conditions (at different levels within the country or the individual's own position in society), all across the life span. Through this approach, the preconception period, pregnancy and early childhood are presented as prime targets for interventions to reduce SIH, the beginning of life constituting a major issue for prevention. These new challenges in social epidemiology raise scientific issues regarding the collaboration between human/social sciences and biology (animal models), raise methodological questions regarding the use of large cohort studies and raise ethical and societal challenges, in particular related to the way causal chains linking the social with the biological could be integrated in public policy, that is to say the way such results may emerge in the context of public debate and/ or be transmitted to the lay public.

We will concentrate on one aspect of the links between early life and health, highly amenable to public health interventions, developed and known as social epigenetics. This approach, sounds a priori as a clash between two terms, evoking on one hand, a new development of an ever more efficient technical and biological medicine and on the other hand, social concerns reminding social medicine of the nineteen century. As a matter of fact, the two worlds are not so far away and a common questioning on these approaches can make an important contribution to the debate on the reduction of social inequalities in health.

Epigenetics: a pathway from environment to biology

Epigenetics refers to any inheritable information during cell division outside the DNA sequence itself[1]. Epigenetic processes are essential to the understanding of gene function and expression. This is a normal process of regulation of gene expression, essential for growth, development and aging of developed organisms. A key point is that this regulation of the genome expression can modify the susceptibility re-

---

[1] Feinberg A. P., "Phenotypic plasticity and the epigenetics of human disease", *Nature*, 2007, 447 (7143): pp. 433 - 440.

garding disease. Epigenetic changes at the molecular level have been involved in the developmental origin of a number of common diseases such as cardiovascular disease, cancer or certain psychiatric pathologies[1][2]. But beyond that, they have been identified as a plausible link between environmental factors, changes in gene expression and susceptibility to disease[3]. One of the basic mechanisms of the epigenetic modifications is methylation of critical regions of genes can affect gene expression, hyper – methylation being usually associated with a reduction of gene expression, while hypomethylation is associated with gene activation. These alterations can be stable and have long – term effects but reversible.

An unfavorable environment may alter biological processes involved in the development of a future disease, especially in case of early exposure. In the 1990s, Barker et al. [4] showed that intrauterine growth restriction was correlated with an increased risk of cardiovascular and metabolic diseases in adulthood, developing the concept of "fetal programming". This concept postulated that environmental conditions occurring during specific windows of development can have long term effects on organogenesis, metabolic and physiological processes. Since then, some epidemiological studies have shown the role of the environment on future health outside the fetal period. One of the major issues of social epidemiology is to understand how the biological, psychosocial and environmental factors are linked and identify pathways through which these factors can influence health.

The life – course approach offers a theoretical framework to better explore these

---

[1] Hochberg Z. et al., "Child Health, Developmental Plasticity, and Epigenetic Programming", *Endocr Rev.* 2011, 32 (2): pp. 159 – 224, .

[2] Szyf M., "The Early Life Environment and the Epigenome", *Biochim Biophys Acta.* 2009, 1790 (9): pp. 878 – 885.

[3] Jirtle R. L., Skinner M. K., "Environmental Epigenomics and Disease Susceptibility", *Nat Rev Genet.* 2007 8 (4): pp. 253 – 262.

[4] Barker D. J., Godfrey K. M., Fall C., Osmond C., Winter P. D., Shaheen S. O., "Relation of Birth Weight and Childhood Respiratory Infection to Adult Lung Function and Death from Chronic Obstructive Airways Disease", *BMJ.* 1991, 303: 671 – 5, 1991.

links. This approach is defined as the study of the long – term effects on health of physical and social exposures occurring from pregnancy, childhood, adolescence to adulthood but also across generations. This area includes the study of biological, behavioral, and psychosocial mechanisms that operate through the entire life of an individual to influence his health. This approach emphasizes the notion of causality chains, "the cause" being then the pragmatic designation of a point in the chain of events on which it is possible to intervene.

These works attempt to study how environments, particularly social and psychological ones, "get under the skin" to explain some parts of the social gradient in morbidity and mortality. An explanation that is still little emphasized lies in a direct biological action of social and psychosocial determinants. This direct biological effect is in accordance with the notion of biological incorporation (embodiment) of the environment in a broad sense (physical, chemical, etc.), and in particular of economic and psychosocial environments, and how it becomes biological[1].

One of the main biological mechanisms used by organisms to adapt to their environment includes stress response systems, notably via the hypothalamic – pituitary system and the sympathetic nervous system, which control the release of stress hormones. Exposed to chronic stress, these systems are in a constant state of "overheating"[2] leading to adverse consequences on the regulation of biological systems.

Epigenetics is then that one of the biological mechanisms by which the body adapts to its environment and ultimately by which the environment "gets under the skin."

Stressful experiences in childhood may then lead to changes in biological systems, involving specifically the hormonal inflammatory and immunological systems,

---

[1] Krieger N., "Embodiment: A Conceptual Glossary for Epidemiology", J Epidemiol *Community Health*, 2005, 59 (5): 350 – 5.

[2] Lutgendorf S. K., Sood A. K., Antoni M. H., "Host Factors and Cancer Progression: Biobehavioral Signaling Pathways and Interventions", *J Clin Oncol.* 2010, 28 (26): pp. 4094 – 4099, 2010.

or epigenetic mechanisms and gene regulation.

A subtle play of interrelations between biological functions and the environment without determinism, ongoing throughout life

Two mechanisms are in fact closely related. The existence of sensitive periods during the lifecourse cannot be easily disentangled to other mechanisms such as a process of accumulation or series of adverse events over life. The way through which biological and behavioral processes interact with psychosocial factors is probably different according to social position.

The fact that childhood prepares the future is an idea commonly accepted, but we also know that individuals can be resilient to life's difficulties. It is known that some individuals can overcome dramatic episodes in their lives. Human development is characterized by periods of rapid growth. Environmental factors are more likely to have an impact on health if exposure occurs during these sensitive periods. The notion of a unique sensitive period is too schematic and it is likely that sensitivity levels are constantly changing according to different physiological systems. These sensitive periods occur most frequently in the first years of life, but continue to occur over time.

Although research works suggest that during this sensitive period a stress exposure could lead to the development of chronic diseases, it is too simple to think that the body plasticity stops and that damage is fixed and will lead to disease. Human beings testify indeed a remarkable physiological adaptability, as brain plasticity, throughout life. The question of how to intervene in individuals with increased susceptibility during certain phases of their development and in specific areas (educational, nutritional) is essential. Children who have faced adversity during childhood should not be labeled or considered in a deterministic view as exposed to developing chronic disease or psychopathology. The great adaptability of live beings is the probably biggest in humans due to their cognitive and socio – emotional capabilities. At an early stage of life, when parental investment is essential, interventions to improve parental investment could prevent maladaptive emotional situations and deregulation of biological systems of child, improving its response to events potentially harmful to health[1]. It is clear that sensitive periods cannot be easily separated from a cumulative effect of exposures, the combination of these two mechanisms during the life-course probably increasing adverse health outcomes. It is noteworthy to remind that there are, beyond childhood, other sensitive periods, whether biological (menopause for example) or social (entering working life, retirement, parenthood ...), justifying thinking biological incorporation of the environment as a continuous process.

Childhood adversity and health in adulthood: an example of epidemiologic results

In a British birth cohort of people born in 1958, the influence on health of intra-family stressful events reported at the age of 7, 11 et16 years including the fact, for children, of having died, divorced, separated, alcoholic parents but also problems of malnutrition or abuse reported by teachers has been studied. A link has been shown between these adversities during childhood, and the occurrence of cancer before the age of 50, regardless of known cancer risk factors, such as tobacco, alcohol and other variables such as socioeconomic environment during childhood, level of education in adulthood, age of first pregnancy or depression. A cancer risk twice higher was observed for women who have been exposed during their childhood to at least two adversities compared to those who did not encounter any adversity[2]. On the same cohort, an association between childhood adversity and premature mortality (before 50 years old) from all causes was also observed. Men who have been exposed to at least two adversities had a 57% higher risk of dying before age 50 than those who did

[1] Mayes L. C., Swain J. E., Leckman J. F., "Parental Attachment Systems: Neural Circuits, Genes, and Experiential Contributions to Parental Engagement", *Clin Neurosci Res.* 2005, 4 (5-6): pp. 301-313.

[2] Kelly-Irving Metal, "Childhood Adversity as a Risk for Cancer: Findings from the 1958 British Birth cohort study", *BMC Public Health* 2013, 13: 767, 2013.

not encounter any adversity. This increase in mortality was 80% in women[1].

Consequences for prevention and reduction of social inequalities in health

This new knowledge on biological incorporation of the environment over time, including sensitive periods such as the early years of life, and the development of epigenetics as a mechanism for this incorporation does not involve and does not allow for conclusions at the individual level. In particular no screening or individual prediction should be expected from this new knowledge which presents in contrast a collective dimension. They show the importance of environmental, physical, chemical and social factors on the development of a number of organs, health, and more newly on gene expressions. Regarding interventions, it is therefore a collective approach, from the first years of life, that is the most appropriate. If we take the example of a particular critical period and a vital organ, the brain, it is certain that the first years of life are a sensitive period. Works led in a UK cohort that studied children during their first five years show that emotional intellectual and personal development, measured through a variety of scores, changes considerably according to parental socioeconomic position. Children whose parents have a higher socioeconomic position maintain an advantage on these tests from the ages of 22 months to 120 months, or will catch up with children born with a high score if they have a low score. In contrast, children whose parents have a low socio – economic position gradually lose their advantage during follow – up if they had a high initial score and will not progress if their initial score was low[2]. The opportunities to improve this cognitive development through the provision of child care arrangements in the early years have been shown. With no effect on children whose mothers had a high level of education, such a provision allows one to compensate for the disadvantage of children whose

---

[1] Kelly – Irving Metal, "Adverse Childhood Experiences and Premature All – cause Mortality", *Eur J. Epidemiol* 2013, 28 (9): pp. 721 – 734, 2013.

[2] Feinstein L. Inequality in the early cognitive development of British children in the 1970 cohort. Economica 2003, 70: pp. 73 – 98.

mothers had a low level of education if they have a form of collective or individual care in addition to the family environment[1].

Conclusions

Epigenetics, as one part of the lifecourse approach in epidemiology, may be an incentive to carry out new types of multifactorial interventions. It emphasizes the need to rethink the role of the early environment in the production of the health status of a population and its distribution. Investing in a favorable environment from the early years is important for health and well – being throughout life. Our ability or the ability of human beings to adapt to their environment also means that interventions are possible throughout life to improve health and well – being: thus it is never too early, but it is also never too late.

[1] Geoffroy M. C. et al., "Closing the Gap in Academic Readiness and Achievement: The Role of Early Childcare", *J. Child Psychol Psychiatry*, 2010, 51: 1359 – 67.

Part Three　Medicine and Law under Opportunities and Challenges

IN NEED FOR A MODERN DAEDALUS? THE CHALLENGING REGULATORY PATH FOR MARKETING GENE THERAPY MEDICINAL PRODUCTS IN CHINA AND EUROPE

A. Mahalatchimy[1], A. – M. Duguet[2], W. Meng[3], H. C. Howard[4], A. Cambon – Thomsen[2], E. Rial – Sebbag[2]

Acknowledgement: This work has been partially funded by REGenableMED, UK ESRC Project ES/L002779/1, in which Aurélie Mahalatchimy is a member

Abstract:

This article highlights the regulatory challenges of the assessment process of Gene Therapy Medicinal Products. It also addresses the puzzling institutional maze that influences the regulatory path for the marketing of these products as shown through various examples in the European Union and in China.

Advanced therapies raise critical questions, especially concerning their pre – market evaluation, which could differ greatly between regions worldwide, and in particular between Europe and China.

After years of hopes and disappointments in the field of gene therapy where failures gave rise to high effervescence widely relayed by the media at the detriment of rare successes, few Gene Therapy Medicinal Products (GTMP), (the use of genes

[1] CNRS, International, Comparative and European Laws – DICE – CERIC – Aix – Marseille University, Toulon University, Pau & Pays Adour University Centre for Global Health Policy, School of Global Studies, University of Sussex, UK.

[2] UMR/INSERM Unit 1027 Université Paul Sabatier Toulouse France.

[3] School of law, Shandong University, 250100 Jinan, China.

[4] Centre for Research Ethics and Bioethics, Uppsala University, SE – 751 22, Uppsala, Sweden Correspondence to: aurelie. mahalatchimy@ gmail. com

to treat or prevent a disease), have ever reached the market. In 2003 Gendicine was approved by the Chinese State Food and Drug Administration (SFDA). It became the first GTMP to be marketed in the world. Its European counterpart, Advexin, as well as Cerepro did not obtain European marketing authorization. Almost ten years later, in October 2012, Glybera was finally granted marketing authorization by the European Commission (Table). After a complex regulatory process, it became the first GTMP to be approved in Western countries. Although approvals of these GTMP have mainly been presented as a huge step forward for patients' treatments and for the development of GTMP in general, no publication has addressed the major challenges present in the regulatory processes as revealed by the examples of these four products both in China and in the European Union (EU).

Table: GTMP regulatory assessment in the world

Regulatory paths for new drug approval

In the EU, the commercialization of GTMP relies on a marketing authorization which involves both the European Medicine Agency (EMA) and the European Commission. As part of a specific regulation[1], the EMA, in charge of the evaluation process of new drugs, has set up a new specific and multidisciplinary committee to assess advanced therapy medicinal products, including GTMP, called the Committee for Advanced Therapies (CAT). In the EU, a regulation is a powerful legal harmonization instrument; it directly applies "as is" in the various EU Member States and unlike EU directives, it does not require a specific transposition into national legal systems. One of the CAT tasks is to formulate draft opinions on the quality, safety and efficacy of GTMP for final opinion by the Committee for Medicinal Products for Human Use (CHMP). Scientific experts have a main role in the assessment of risks

[1] Regulation (EC) N°1394/2007 of the European Parliament and of the Council on advanced therapy medicinal products and amending Directive 2001/83/EC and Regulation (EC) N°726/2004 OJ L324, 10/12/2007, p. 121.

and benefits linked to the use of the products. The final decision of granting marketing authorization is in the remit of the European Commission. This political body takes the final decision integrating other dimensions above and beyond the scientific evaluation.

In China, the SFDA is the only competent organization to evaluate the safety, efficacy and quality of drugs for marketing, and has the power to decide whether or not a product is approved[1]. A registration approval for the manufacture of the new drug is needed for market access. It involves obtaining the following approvals by the SFDA: new drug certificate, new drug registration certificate and drug Good Manufacturing Practice certificate[2]. Within the SFDA, the Center of Drug Evaluation which organizes panels of experts is in charge of the technical assessment of applications but the final decision is taken by the SFDA. Since February 22nd, 2013, the SFDA has been restructured and renamed the China Food and Drug Administration (CFDA). It became a ministerial – level agency to improve food and drug safety[3], and included several proposals "to boost confidence in the drug review and approval process as well as help promote regulatory oversight"[4], as deemed necessary[5].

Assessment process of GTMP

Among the GTMP that have been assessed at this time, we target our analysis on those that aimed to reach the European or Chinese markets (although it should be noted that Rexin reached the Philippines' market in 2007 to treat all chemotherapy

[1] Article 3, SFDA Order N° 28, Provisions for drug registration, 01/10/2007.

[2] H. Yin, "Regulations and Procedures for New Drug Evaluation and Approval in China", *Hum. Gene Ther*, 17, pp. 970 – 974 (2006).

[3] China State sponsored news – site "xinhuanet": Bi Mingxin, China gets stronger food, drug regulator, March 22nd, 2013: http://news.xinhuanet.com/english/china/2013 – 03/22/c_ 132253914.htm (Access date 6th October, 2014).

[4] H. Jia, "China overhauls Drug Regulation Agency", *Nature Biotechnol*, 31, 5: 375 (2013).

[5] Gareth Macdonald, SFDA aims to stimulate R&D, accelerate approvals and improve distribution practices, 5 March 2013, in – Pharma Technologist. Com website: http://www.in – pharmatechnologist.com/Regulatory – Safety/SFDA – Aims – to – Stimulate – R – D – Accelerate – Approvals – and – Improve – Distribution – Practices (Access date 6th October, 2014).

resistant solid tumors).

Gendicine and Oncorine

Since 1999, the development of Gendicine and Oncorine has been financially supported through several Chinese government plans. For Gendicine, the SFDA delivered the Biological type I New Drug Certificate in October 2001, the Production Permit in January 2004, and the Drug GMP Certificate in March 2004. Gendicine also obtained the National Key New Product Certificate in 2005 issued jointly by the Ministry of Science & Technology, the Ministry of Commerce, the General Administration of Quality Supervision, Inspection & Quarantine and the Bureau of Environmental Protection. Oncorine obtained the Biological type I New Drug Certificate in November 2005, the Drug GMP Certificate and the Production Permit in 2006. Since 2004 and 2006, the Production Permits of Gendicine and Oncorine have been regularly renewed.

Glybera

Amsterdam Molecular Therapeutics B. V. applied for marketing authorization at the EMA in October 2009. Before the process was completed the applicant changed from Amsterdam Molecular Therapeutics to UniQure Biopharma. In June 2011, the CAT and CHMP adopted negative opinions because of a lack of evidence of long - lasting benefit in the patients studied and a lack of reduction in the rate of pancreatitis (the clinically relevant endpoint). Additionally they had concerns over the risks linked to the associated immunosuppressive treatment. The company requested a re - examination. After several procedural steps, the CAT adopted in October 2011 a positive opinion conditioned by the restriction of the indication. But, the CHMP maintained its previous negative opinion concluding that benefits did not outweigh the risks. In January 2012, the European Commission requested a re - evaluation in a restricted group of patients with severe or multiple pancreatitis episodes. In April 2012, the CHMP gave a third negative opinion. However, for procedural reasons it was invalidated and a new examination was performed; it was conducted in June

2012 and resulted in a final positive opinion. However, 15 CHMP members (out of 32) had a divergent opinion and considered that efficacy and safety had not been sufficiently demonstrated. Finally, the European Commission granted a marketing authorization under exceptional circumstances to Glybera; these include specific obligations for post – authorization measures (such as the setting up of a long – term surveillance of patients) to be conducted in a specified timeframe[1].

Advexin and Cerepro

Both Gendux Molecular Ltd for Advexin and Ark Therapeutics for Cerepro withdrew their applications for marketing authorization at the EMA. The former specified that the company's marketing strategy had changed[2] while the latter admitted it was not able to provide meaningful evidence of benefits compared to the risks[3]. At that time, the CHMP was about to give a negative provisional opinion for Advexin as it had concerns about the lack of evidence regarding the benefits and safety of the product for patients, for people in close contact with them and for the environment[4]. Regarding Cerepro, the CHMP had already delivered a negative opinion. Following a request of re – examination by the company, the CHMP was about to give a new negative opinion due to the lack of effectiveness which caused an emphasis on the concerns regarding side effects[5].

---

[1] Glybera European Public Assessment Report, European Medicines Agency (EMA/882900/2011, 2012; http: //www. ema. europa. eu/docs/en_ GB/document_ library/EPAR_ –_ Public_ assessment_ report/human/002145/WC500135476. pdf).

[2] Letter of withdrawal for Advexin (Gendux Molecular Limited, 2008; http: //www. emea. europa. eu/docs/en_ GB/document_ library/Other/2010/01/WC500063082. pdf).

[3] Letter of withdrawal for Cerepro (Ark Therapeutics, 2009; http: //www. ema. europa. eu/docs/en_ GB/document_ library/Other/human/001103/WC500076159. pdf).

[4] Withdrawal Assessment Report For Advexin, European Medicines Agency (EMEA/692328/2008; http: //www. emea. europa. eu/docs/en_ GB/document_ library/Application_ withdrawal_ assessment_ report/2010/01/WC500063080. pdf).

[5] Withdrawal Assessment Report For Cerepro, European Medicines Agency (EMA/CHMP/798830/2009; http: //www. ema. europa. eu/docs/en_ GB/document_ library/Application_ withdrawal_ assessment_ report/2011/02/WC500101545. pdf).

All five of these decisions relied on scientific evaluations which resulted in concerns that were greater regarding the drugs' efficacy than their safety. These decisions, both in Europe and in China, were also largely influenced by other types of variables, such as economic, political and ethical considerations.

The scientific evaluation of GTMP: what prevails efficacy or safety?

Both in China and in the EU, the scientific evaluation of GTMP based on the assessment of the quality, safety and efficacy of the product is difficult even though experts are involved in the process. Based on the media focus on negative outcomes of the use of GTMPs, one could think that the most important challenge of GTMP meeting evaluation criteria would be their safety, but in fact, showing the efficacy of these products has also been shown to be problematic. The approval of Gendicine has been criticized by Western countries as efficacy was based on tumor shrinkage rather than extension of patient lifetime[1]. A greater emphasis may have been placed on safety than on efficacy. Meanwhile safety and lack of transparency on studies' results have also been challenged. Moreover, in the EU, Glybera, Advexin and Cerepro suffered from several negative (provisional) opinions at the EMA's relevant committees also regarding the lack of proven efficacy. For all these GTMP, the concerns raised have been linked to the lack of data due to the low number of patients involved in clinical trials. Given that Advexin, Cerepro and Glybera, have been designated as orphan drugs in the EU, it is not surprising that they have this problem. The questioning on efficacy of GTMP challenged the benefits/risks balance: should GTMP considered safe be authorized if their efficacy is limited? Interestingly in the USA, the Advexin application file was refused as not complete while Glybera is now approaching the Food and Drug Administration. Will the acceptability threshold differ according to the national agencies? Forthcoming Chinese reforms aim notably to empha-

[1] M. L. Edelstein, M. R. Abedi, J. Winxon, "Gene therapy Clinical Trials Worldwide to 2007 – an Update", *The Journal of Gene Medicine*, 9: pp. 833 – 842, DOI: 10.1002/jgm.1100 (2007).

size the clinical value of innovative drugs, to encourage the development of drugs that answer clinical needs and have a better therapeutic effect, through a speed - up review[1]. Actions such as EMA specific guidance on the scientific assessment criteria for GTMP would also be useful in the EU.

Other influences beyond the scientific evaluation of GTMP

When scientific assessments do not give a clear answer, other dimensions are taken into account and this is particularly true for GTMP. First, the marketing authorization is a political decision deemed to represent what is acceptable for society. Both in China and in the EU, political decisions played a major role in the marketing authorization of Gendicine, Oncorine and Glybera. Second, economics is a main dimension. As "the global gene therapy industry has the potential to become a multi - million dollar industry by the end of 2017 as new products […] may enter the market to boost the growth. "[2], the race is launched for leadership in this field. Whereas China authorized the first gene therapy in the world, the EU adopted a specific regulation aiming to enhance competitiveness of European enterprises developing advanced therapies. The authorization of Glybera demonstrates that a marketing authorization can be obtained for GTMP in the EU[3]. But going through the whole regulatory process is complex, costly and time consuming. When withdrawing its application for Advexin, Gendux Molecular Ltd gave the argument of limited resources to face the regulatory demand. After several negative opinions of the CHMP, Amsterdam Molecular Therapeutics stopped investment in Glybera and transferred gene therapy assets to Uniqure[4]. Moreover, the price of GTMP will be a high burden. Glybera "could cost as much as $ 1. 6 million for the single injection necessary to

---

[1] CFDA [Ideas on deepening the reform of pharmaceutical review and approval and further encourage innovation], 26 February, 2013 (In Chinese; http: //www. sfda. gov. cn/WS01/CL0051/78609. html).

[2] Global Gene Therapy Market Anlaysis, *Global Biological Therapy Industry* (2013).

[3] Geoff Watts, Gene therapy to be authorised for first time in EU, BMJ 2012: 345.

[4] N. Moran, "First Gene Therapy Nears Landmark European Market Authorization", *Nature Biotechnol.* 30, pp. 807 - 809 (2012).

confer lifetime therapy", a pricing that would be inappropriate for more common diseases[1]. Finally, ethical considerations may also influence marketing authorization of GTMP. Medical tourism is an issue for patients' safety and for the environment. Despite the lack of clear data on its real extent, it seems that 300 to 600 patients travelled to China for Gendicine[2]. This very complex issue gives rise to large bioethical discussions that will not be developed in this paper notably as many countries, including the emerging ones, target the global health care market.

It is clear that approvals do not rely only on scientific assessment and include other factors. If a country approves a GTMP, patients will benefit from new treatments which could lead toward less medical tourism out of that country. However, the question remains: will these patients really have access to effective drugs? Regarding the efficacy criterion, collaboration between EMA and CFDA is necessary for the benefits of patients. For drugs focused on rare diseases, it would permit to have a higher number of patients to assess the therapeutic effect. Moreover, a lot of date is generated from Gendicine in China that would be useful worldwide for the future of gene therapy if transparency is improved. Even though the EMA supports the European Commission's collaboration with China, regulatory agreements are not yet developed at a scale and place that would match the level of scientific collaboration. To achieve translational medicine, all dimensions at stake including the regulatory ones needs to be collaboratively developed[3].

Acknowledgements: This work has been supported by CAI YUANPEI Program 2012 – 2014 on Medical Law and Bioethics (N°28007 UF) co – led by E. Rial –

---

[1] J. Whalen, Gene – therapy Approval Marks Major milestone, *Wall Street Journal* (November 3, 2012), p. B3. Cited in Christopher H. Evansa, Steven C. Ghivizzanib, Paul D. Robbins, "Arthritis Gene Therapy and Its Tortuous Path into the Clinic", *Translational Research*, Volume 161, Issue 4, pp. 205 – 216.

[2] L. C. M. Kaptein, Y. Li and G. Wagemaker, Gene Therapy in China from a Dutch Perspectives (report commissioned by the Netherlands Commission on Genetic Modification, 2010).

[3] E. Meslin, A. Blasimme, A. Cambon – Thomsen Mapping the Translational Science Policy Valley of Death, *Clin Transl Med.* Jul 27, 2 (1): 14 (2013).

Sebbag (France) and Q. Yanping (China) and XU GUANGQI Program 2012 on Patients' rights and access to genetic testing (N°27974 QH) co – led by A. Cambon – Thomsen (France) and H. Man (China).

BIOSIMILAR OR BIO - GENERIC IN E. U, FRANCE, AND CHINA EXCLUSIVE RIGHTS[1] IN PHARMACEUTICAL INDUSTRY

LI Mou[2]

Abstract:

Is Biosimilar the same as generic, a copy to the brand name drug or is it simply just another drug with bio - similarity to the brand name drug? This paper is trying to discuss this specific question in exploring exclusive rights in the pharmaceutical industry.

In many countries, policies to reduce pharmaceutical costs have included incentives and regulations to encourage the substitution of cheaper generic drugs to the prescription of their usually more expensive brand name counterparts. Results of policies to encourage prescription of generic drugs however widely differ from one country to another.

Patent issue are important and essential to the pharmaceutical industry.

Exclusive rights in the pharmaceutical industry are important due to the fact that it is an industry that is highly innovative and technologically advanced.

The rationality for intellectual property and for public health are controversial. Both sides are trying to reach their goals either different strategies in patent

[1] In Anglo - Saxon law, an exclusive right is a de facto, non - tangible prerogative existing in law (that is, the power or, in a wider sense, right) to perform an action or acquire a benefit and to permit or deny others the right to perform the same action or to acquire the same benefit. A "prerogative" is in effect an exclusive right. The term is restricted for use for official state or sovereign (i. e., constitutional) powers. Exclusive rights are a form of monopoly.

[2] Ph. D in Law at Toulouse University, CSC and Toulouse University sponsored scholarship, doctoral thesis is under the supervision of professor J. Larrieu, and professor A. M Duguet.

drugs and their prices, or setting restriction on drug circulation. One of the results of this game between two parties, is the result of copying drugs from patents or blockbuster drugs. Technology has improved and grown rapidly, it has gone from small molecule drugs to biotechnology oriented drugs. Thus, the issue of coping also changes from generics to biosimilars.

However, biosimilar is discussed worldwide, there is no single procedure that is agreed among nations. It can be seen as reproducing of the generic pathway, but in some countries it is considered no difference to the new drug. The differences between small molecule drugs and biologics are the basis for the differences in the regulatory standards for their respective abbreviated pathways.

This paper is trying to make it clear and easy to understand picture of the rationality behind each innovative strategy that pharmaceutical industry is working on, and due to what reason generics and biosimilars these two different drugs are most of the time taken into the discussion of national strategies as one.

Keywords: Exclusive rights, Patent, Biosimilar, Generic, Pharmaceutic, Biotechnology

Introduction

The patent system is designed to balance the interests of exclusive rights and the disclosure of invention. A patent is an exclusive right granted by law in a three dimesional[1] protection system. In return for exclusive rights, the applicant is obliged to disclose the invention to the public in a manner that enables others to replicate the invention.

In relation to property, an exclusive right[2] will, for the most part, arise

---

[1] A patent is an exclusive right granted by law to applicants to make use of and exploit their inventions for a limited period of time, generally 20 years from filing in a given territory with listed rights to perform and to prevent others to perform.

[2] For example, in property law, a person may prohibit others from entering and using their land, or from taking their personal possessions. However, an exclusive right is not necessarily absolute, as an easement may allow a certain level of public access to private land.

when something tangible is acquired; as a result, others are prevented from exercising control of that thing.

Exclusive rights are forms of monopoly to perform an action or acquire a benefit and to permit or deny others the right to perform the same action or to acquire the same benefit. Exclusive rights can be established by law or by contractual obligation, but the scope of enforceability will depend upon the extent to which others are bound by the instrument establishing the exclusive right. Exclusive rights may be granted in property law, copyright law, patent law, in relation to public utilities, or, in some jurisdictions, in other sui generis legislation. The notion exclusive is derived from the concepts of property and ownership[1][2].

Patents on pharmaceutical drugs provide a very strong economic incentive[3] for research. Thus, patent or brand named drug companies are seeking for their profit after investing, but on the other side pharmaceutical companies are trying to earn their profit once the intellectual protection period is over.

I. Provides with exclusive rights in the patented invention

1. 1. 1. Right to exclude

Most governments recognize a bundle of exclusive rights in relation to works of authorship, inventions, and identifications of origin. These rights are sometimes spoken of under the umbrella term "intellectual property". Patents are arguably the strongest form of intellectual property rights. A patent typically is issued by a government agency.

A patent affords a right to exclude[4] others, not an affirmative right to practice

[1] Privately granted rights, created by contract, may occasionally appear very similar to exclusive rights, but are only enforceable against the grantee, and not the world at large.

[2] A patent affords a right to exclude others, not an affirmative right to practice what you've patented.

[3] The R&D requirements of creating new drugs are significan't, and the potential payoff that a patent grants encourages companies to make the necessary investments.

[4] Professor Robert Merges best explains that the exclusion concept of patents is "necessitated by the existence of blocking patents", because otherwise an overlapping patent would necessarily result in an illegitimate restriction of another property owner's "affirmative right to actually carry into practice a particular invention".

which one has patented. Patents are defined differently from real property—patents must secure only a negative right to exclude. It is "elementary" that "a patent grants only the right to exclude others and confers no right on its holder to make, use, or sell" the invention.

It confers to the inventor the sole right to exclude others from economically exploiting the innovation for a limited time (20 years from the date of filing, for most countries). To be patentable[1][2], an innovation must be novel[3][4] that is to say not constituting part of the prior art. The innovation must involve an inventive step[5][6], and it must be useful[7][8].

The exclusive rights provided by a patent are founded upon the claims, though they are not necessarily limited to them. Patents confer the right to exclude others from making, using, selling, offering to sell, or importing into the states of the patented Invention. The grant of a patent provides the inventor with a means to capture the returns to his invention through exclusive rights on its practice for 20 years from the date of filing. The patent holder has the legal right to exclude others from commercially exploiting this invention for the duration of a certain period.

Patent ownership is an incentive to innovation, which is the basis for the technological advancement that contributes to economic growth. It is through the commercialization and use of new products and processes that productivity gains are made and the scope and quality of goods and services are expanded. The award of a patent is intended to stimulate the investment necessary to develop an idea and bring it to the marketplace embodied in a product or process. The patent title provides the re-

---

[1] Art. 27. 1 of TRIPS agreement.

[2] Art. 22 Chinese Patent law.

[3] Art. 54 European patent convention.

[4] Article L. 611 – 11 CPI in French intellectual property code.

[5] Art. 56 European patent convention.

[6] Article L. 611 – 14 CPI in French intellectual property code.

[7] Art. 57 European patent convention states as industrial application.

[8] Article L. 611 – 15 CPI in French intellectual property code as industrial application.

cipient with a limited - time monopoly over the application of his discovery in exchange for the public dissemination of information contained in the patent application. This is intended to permit the inventor to receive a return on the expenditure of resources leading to the discovery but does not guarantee that the patent will generate commercial benefits. The requirement for publication of the patent is expected to stimulate additional innovation and other creative means to meet similar and expanded demands in the marketplace.

Innovation typically is knowledge - driven and is based on the application of knowledge, whether it is scientific, technical, experiential, or intuitive. Innovation also produces new knowledge. One characteristic of knowledge that underlies the patent system is that it is a "public good," a good that is not exhausted when it is being used.

1.1.2. Incentives to Innovate

The logic behind patents is a form of rights compromised by individual and social benefits. Society agrees to grant patent - holders a temporary monopoly, thus allowing them to set prices at an artificially high level, on the understanding that doing so will provide a strong incentive[1] for innovation.

Moreover, patent rights would be "enjoyable without discrimination as to the place of invention, the field of technology and whether products are imported or locally produced."

1.2. Exclusion concept of patents

The evidence of the validity of the exclusion concept of patents is the phenome-

[1] In the case of an industry constantly creating new products, such as the pharmaceutical industry, the assumption is that much of the increased profits will be fuelled back into research, resulting in new treatments for diseases being discovered.

na of blocking patents [1]. A blocking patent [2] exists when two separate patents cover aspects of the same invention, and thus each patentee can exercise his right to exclude the other patentee from using his respective contribution to this invention.

Such situations are quite common, as inventive activity often builds on earlier innovation, and thus prior inventors are able to exclude [3] follow - on commercial applications of their inventions. On the other hand, a public regulation may impose on a property owner a different set of substantive restrictions that achieve normative goals that are distinct from those already at work within the pre - existing set of property entitlements. In such a situation, property owners are now subject to two sets of legal regimes—a public regulatory regime and a private property regime—and each regime represents differing substantive requirements for the use of the property.

1. 3. Territorial exclusivity

The rights provided by patents are ordinarily effective only in the certain [4] region/ nation.

II. Exclusivity in pharmaceutical industry

The utility of patents to companies varies among industrial sectors. Patents are critically perceived in the drug and chemical industries. The rational behind the critic

---

[1] Adam Mossoff, "Patent as Property: Conceptualizing the Exclusive Right (s) in Patent Law", *Select Works*, September 2008.

[2] A obtains a patent on a new product, such as a new drug. Several years later, B discovers a new process for using A's drug, and this discovery constitutes a patentable invention itself (the process is novel, non-obvious and has utility). The result is that two patents held by A and B cover overlapping aspects of the same invention: (i) the drug and (ii) a particular process for using the drug. A can thus exercise his right to exclude B from using his patent drug in commercially exploiting his new process, regardless of B's inventive act in discovering a new use for A's drug. In this situation, A has a "blocking patent," because he can block B's use of his own patented process. (Concomitantly, B can also exclude A from using his process, but A has the greater scope of exclusivity here, because he has a prior claim in the product, which he can continue to use as long as he avoids B's patented process).

[3] The qualification of novelty.

[4] They generally provide no protection against acts occurring in foreign countries. Individuals must obtain patent protection in each nation where they wish to guard against unauthorized use of their inventions.

can be seen as it may reflect the nature of R&D being performed in these sectors[1], where the resulting patents are more detailed in their claims and therefore easier to defend.

Exclusive rights in the pharmaceutical industry are important due to the fact that it is an industry that is highly innovative and technologically advanced and has consistently maintained a competitive[2] edge in international markets.

2. 1. 1. Patent and market approval

Both the patent office and the administration entity[3] have a role[4] to play in the pharmaceutical industry.

The grant of a patent does not provide its proprietor with the affirmative right to market the patented innovative drug, however. For many products of the pharmaceutical industry, the national authority must approve[5] the product for sale to consumers.

National law generally requires that pharmaceutical manufacturers show their products are safe and effective[6] in order to commercialize these products. The issuance of a patent and a marketing approval are distinct events that depend upon different criteria[7].

---

[1] In certain industries, patents significantly raise the costs incurred by non - patent holders wishing to use the idea or invent around the patent - an estimated 40% in the pharmaceutical sector, 30% for major new chemical products, and 25% for typical chemical goods - and are thus viewed as important. However, in other industries, patents have much smaller impact on the costs associated with imitation (e. g. in the 7% - 15% range for electronics), and may be considered less successful in protecting resource investments.

[2] Highly innovative.

[3] EMA is the administration authority at European level, SFDA is the state food administration in China, and French is ANSM.

[4] For example, the fact that an inventor obtains a patent on a pharmaceutical compound does not allow him to market this medication to others. Approval of the appropriate food and drug authorities must also be obtained.

[5] Special attention must be followed as obtaining the approval for AMM, each nation has its own procedures and caution on granting the approval as well as in different stages for the approval.

[6] The safe and effective procedure are taking into considerations at European level for the similarity interchangeable criteria in biosimilars.

[7] This part will not be developed in this paper but it is in the studies of my research.

2. 1. 2. The independence of patent and Marketing Authorisation Application (MAA)

As a result of the independence of patent ownership and marketing approval, the pharmaceutical industry must account for both. In order to avoid any foreseeable fear of civil or criminal liability for a drug manufactured, an enterprise must both obtain a certain approval and consider whether that drug has been patented. Most of the time the enterprise, which owns the patent on a drug, is the first to be awarded marketing approval. Sometimes the enterprise that has been awarded marketing approval and the patent owner are separate[1] entities, however. In this latter case, the patentee may commence infringement litigation against the approved drug manufacturer.

2. 1. 3. Compulsory licensing[2] and exclusive rights

The compulsory licensing on drugs is a tough question to discuss, as for drugs especially innovative and effective drugs should be made as widely available as possible to prevent a certain disease from spreading, which is the logic for national[3] authorities granting the compulsory license under specific circumstances some times can be recognized flexibilities. [4] But compulsory licensing can be an obstacle for further innovation, and will discourage pharmaceutical companies making a bigger contribution to upgrade or advance the technology.

---

[1] This is where within the same drug the conflict of exclusive rights arose.

[2] Arguably the most controversial aspect of the World Trade Organization's Agreement on Trade – related Aspects of International Property Rights (TRIPS) is over the issue of patents for pharmaceutical drugs. To their proponents, patent rights are essential to encourage innovation, as the virtual monopoly they allow firms to extract greater profits from drugs that they invent. Opponents, however, point out that patents result in higher prices, making essential medicines less affordable.

[3] South Africa, Brazil.

[4] Such as Article 31, which establishes the procedures by which a compulsory license may be granted. According to 31 (f), a compulsory license must be "predominantly for the supply of the domestic market of the Member authorizing such use.".

What can be more dangerous to public health [1] of the entire human race? Today's problem is different from decades ago, that a steady wave of the diseases that come with age, not an out - of - control virus. It requires by varying drugs prices or tailored drug economy that pharmaceutical companies can pep up their profits at the same time as expanding [2] their markets, making both shareholders and the sick ones better off. Access to drugs and health care services is a real problem. Patents on pharmaceuticals, through their impact on higher drug prices, may contribute to the problem, although they may not be the main stumbling block.

Differential pricing—making the costs of medicine cheaper for people in developing countries than in developed countries—is a solution to the issues of health access raised by patents, but this is an imperfect solution. It did not solve the conflicts of interests between exclusive rights that the enterprises are trying to remain their net profit obtained by patent and health access to the public as national strategy. The argument makes economic sense, as developing world markets provide minimal incentive for research: "The total market of the poorest countries…is on the order of 1 percent of the global pharmaceutical market." Moreover, the logic of patents is that they allow prices to be artificially high on the understanding that increased pro? ts

[1] TRIPS figured prominently in the Doha WTO ministerial meeting in November 2001, and resulted in the Doha Declaration on the TRIPS Agreement and Public Health. The concerted effort of developing countries, assisted by the emerging influence of non - government organizations (NGOs), brought before the public health problems afflicting many of them, especially those associated with HIV/AIDS, tuberculosis, malaria and other epidemics. The declaration stressed that TRIPS "… does not and should not prevent members from taking measures to protect public health." Specifically, the declaration (a) recognized that compulsory licensing could be used to procure critical drugs, at each member's discretion; (b) acknowledged that each member is free to adopt the desired mode of exhaustion of IPRs (which bears on whether parallel imports are allowed into a country or not); (c) recognized that developing countries with insufficient drug manufacturing abilities would face difficulties in taking advantage of compulsory licensing, and thus instructed the Council for TRIPS to find a solution to this problem; and (d) extended until January 2016 the deadline for LDCs to implement IPR protection for pharmaceuticals and test data. The problem addressed in point (c) above arises because TRIPS stipulates that compulsory licenses may be used primarily to supply the domestic market.

[2] Some companies are trying this. Roche, a Swiss company, has created new brands and packaging for lower - priced drugs in India and Egypt.

will be fuelled back into research.

2. 1. 4. Exhaustion of Rights and Parallel Imports

TRIPS agreement has explicitly leave the exhaustion of rights and parallel imports to individual Member States to decide[1] either to rely on the doctrine or to implement a national principle. The main difference can be seen at pricing, as national exhaustion allows innovators to price the product differently in different markets, whereas international exhaustion allows arbitrage across markets in order to prevent differentiation of pricing in market. Each individual country may have a unilateral incentive to deviate from this strategy because it could benefit from parallel imports from a market with lower prices. And this will directly fall into the discussion of competition law, which is out of the scope of this article.

2. 1. 5. Conclusion

The patent system is therefore a relatively important player in the pharmaceutical context, but the rationality for intellectual property and for human rights is controversial, thus the debate on consistency for IP and primacy of human rights exist.

National authorities and non - governmental entities are trying to find a balance, in order to solve the conflict. Unfortunately, as what has been studied at national level, to make the protection of patented drugs more sophisticated and restrictive does not really make better off for the public health, and nothing better or we can say in some points it is much more worse at international stage, all the constraints of non - supranational power explicitly show that the patented drugs are not willing to compensate their profits to welfare. Thus, all the means on pricing strategies have to be negotiated.

[1] Efforts in the Uruguay Round to find a unified stance failed, and TRIPS explicitly leaves it to individual countries to decide whether they want to rely on the "international exhaustion" doctrine or whether they want to implement a "national exhaustion" principle. Under the latter, the right of the IPR holder on the product expires with the first sale in that jurisdiction, but the IPR holder retains the right to exclude (parallel) imports and exports in that region. Under the former, the right of the IPR holder expires with the first sale anywhere.

2.2. Industrial strategies: copy or innovation

Possible solutions and awareness of these shortcomings between patented drugs and protection on public health have contributed to the emergence of a number of alternative approaches that attempt to solve in the research system within the pharmaceutical industry. Many of these initiatives[1], however, came up only recently and have not yet been evaluated[2] for effectiveness.

The truth beneath is that the big pharmaceutical companies are trying whatever they can protect the intellectual property (IP) that drugs represent, and the public health policies are putting the moral weight on pharmaceutical companies and patented drugs as people have a right[3] to health care[4]. And thus, the industry itself has to find a pathway to direct things out.

In general, in contemporary pharmaceutical industry, there are two trends[5]: either maintains their competent drugs in the form they want, or taking a new bridge to continue the journey.

The generics came into sight as more and more brand name drugs or the blockbuster drugs are going to expire their exclusive right granted by patent in chemical oriented drugs industry. Biosimilars are the most talked about over the globe nowadays as they are seen as the reproduce of generics in sunrise industry – biotechnology

---

[1] Generics are seen as the copies of patented drugs or so called brand name drugs, and biosimilars are increased in biotechnology generated and are similar to the pioneer drugs pharmaceutical industry.

[2] Under the Trade Related Aspects of Intellectual Property Rights, a deal signed in 1994, governments can allow a generic drug maker to produce a patented medicine. Patents on drugs are in the interests of the sick as well as the industry. Protection should not be weakened of all the goods and services traded in the market economy.

[3] Fundamental human rights vs. intellectual property rights are hot topic nowadays at UN level.

[4] These criticisms reached a crescendo more than a decade ago at the peak of the HIV plague. When South Africa's government sought to legalise the import of cheap generic copies of patented AIDS drugs, pharmaceutical companies took it to court. The case earned the nickname "Big Pharma v. Nelson Mandela". It was a low point for the industry, which wisely backed down.

[5] Second therapeutic application, and secondary drug or second generic of the patent drug.

pharmaceuticals. But they are not the same in almost every sense[1] that is crucial to have the market approval.

2. 2. 1. Competition in generics and brand name drugs

Some experts argue that the expiration of patents and the desire to generate new replacement drugs, not the extension of patent ownership, is the stimulus to innovation. The portions of the legislation that have accelerated the introduction of generic products have affected the brand name firms in various ways that may or may not influence innovation in the industry.

2. 2. 2. Strategies for the brand name drugs

Unlike what said to be expected, the price of the brand name drugs did not follow, but in fact increased after patent expiration, due to their strategic marketing. They are looking forward for a second usage for the patent drug. Despite competition from generics that have appreciably lower prices, the prices for brand name drugs often increase[2] after patent expiration. At the same time, generic prices for the comparable drugs fall.

Such price increases are based on the recognition that when generic substitutes are available, the market bifurcates. Price - insensitive consumers will pay more for a brand name while consumers that respond to price will buy the generic.

---

[1] Difference is distinctive between generics and biosimilars.

To be approved, generic drug applicants must demonstrate "bio equivalency" in addition to provide information regarding the labelling, chemistry components, manufacturing, and quality of its drug product. A generic drug is bio equivalent when the rate and extent of absorption of the drug that is to say its amiability are not significantly different from those of the reference drug.

A biosimilar product is defined as one that is "highly similar" to the pioneer product notwithstanding minor differences in clinically inactive components and where there are no clinically meaningful differences in terms of safety, purity and potency. And it requires the study and proves of "interchangeable" criteria.

[2] Malinowski and Vernon found that innovator drug prices continued to increase at the same rate as before the introduction of generics even as market shares declined.

"Maintaining and even raising the price of the brand - name product on the theory that the demand for it was more inelastic than the demand for the price - sensitive segment; they have embarked on a new aggressive strategy designed to serve the brand - loyal segment and capture a substantial share of the generic market." ——Contrary to expectations.

To protect their market share, brand name companies focus on developing brand loyalty. They are also in the process of re – patenting their blockbuster drugs under the second therapeutic application regime.

2. 2. 3. Biotechnology & biosimilar

The ownership of intellectual property is particularly important to biotechnology companies. Biotechnology drugs may be similar in their chemical or biological make – up but test differently in clinical trials. And because of this difference, biosimilar is totally different from the practice of generics. EMA[1] has established a guideline on biosimilars, but still it did not give a certain procedures for enterprises. At European level, biosimilar applications are examined under safety and efficacy principles, and it has to be looked through case by case, that is to same there is no standard solution – result foreseeable when an enterprise applies for the approval of a biosimilar. The second essential criterion at European level is the similarity between pioneer drug and the similar biologic, and its interchange ability in between. While on the other side of the continent, in China, biosimilar[2] has no prejudice over the other newly patent drugs, has to go through new drugs approval procedures. That means the enterprise has to go through all the stages and clinical trails and to wait for a long time before the biosimilar enter into the market. Neither the process of the approval is favourable for start – up enterprises, nor for big pharmaceutical companies that are not in favour to manufacture biotechnology drugs are not giving any interests at all to biosimilars.

Based on these factors, some in the industry maintain there is a need to create regulations similar to those in the original legislation for biologics in order to develop a generic sector such that exists in pharmaceuticals. As biologics itself is very com-

---

[1] Biosimilar medicines, http: //www. ema. europa. eu/ema/index. jsp? curl = pages/special_ topics/document_ listing/document_ listing_ 000318. jsp.

[2] 2013 report on biosimilar in China, http: //doc. mbalib. com/view/3cbb94bffbc204b3b36901144abe9570. html

plicated, to set up a standard formula like the ones used in other technology is not a good idea, and especially this standard will be one for drugs, which has great influence on public health.

How do patent term extensions and market exclusivity provisions encourage and/or facilitate activities by firms that might not foster innovation? For example, the law provides the opportunity to extend market exclusivity by listing patents, and separate different types of usage of patent. But will this be too sophisticated as to put into the patent code? Legislation should always be neutral and not exhaustive. And for drugs, and other sensitive patents to have a general legislation are more rational than to have a very specific one.

Ⅲ. Difference of generic and biosimilar in science and in regulatory standard

3. 1. Why not bio - generics but biosimilar? The differences between small molecule drugs and biologics are the basis for the differences in the regulatory standards for their respective abbreviated pathways.

Biosimilars[1] sometimes also referred as "follow - on biologics". Biologic drugs, however, are much more complex than small molecule drugs. These drugs are proteins that must be produced by a living cell——they cannot be chemically synthesized in the lab by following a standard set of procedures[2].

A generic drug is essentially a copy version of small molecule brand name drug that can be synthesized in the lab by following standardized, pre - defined procedures. Using well - established analytic techniques, the generic version of a small molecule drug can be demonstrated to be chemically and structurally identical to the

[1] Biosimilar is the term used for a biologic drug that is produced using a different cell line, master cell bank, and/or different process than the one that originally produced the product.

[2] The cell makes these proteins by following a recipe provided by a short sequence of DNA - a gene - that is inserted into the cell. Here's the catch: even if two different cells are provided the exact same recipe, the final product may be slightly different. This may happen even if the two cells are of the same type - very slight environmental differences can have a profound effect on how a given cell follows a particular recipe.

innovator drug.

Complicating matters further for biosimilar products is the fact that because biologic drugs are structurally much more complex than their small molecule counterparts, it is not currently possible to demonstrate conclusively that a biosimilar drug is in fact identical to the original biologic drug. Thus, since we know that there is a high likelihood that a biosimilar drug is not identical to the original biologic, and we have no way of precisely measuring whatever differences may exist, the term "biosimilar" is used rather than "generic", which implies identity.

3. 2. Regulatory standards for generics

A Generic is a chemically derived medicinal product, usually a simple, homogeneous small molecule, whose bioequivalence must be demonstrated.

In many countries, policies to reduce pharmaceutical costs have included incentives and regulations to encourage the substitution of cheaper generic drugs to the prescription of their usually more expensive brand name counterparts. Results of policies to encourage prescription of generic drugs however widely differ from one country to another.

France presents a situation in which a generic drug market depends basically on whether physicians choose a prescription method that can lead to the delivery of these drugs and on patients' willingness to accept them.

On the other hand, China has taken generic as an innovative drug in their pharmaceutical development. On 22 February last year, the SFDA in China has released "suggestions on encouraging innovative drugs", it is said to improve the administration approval on innovative drugs and the valuable therapeutic generics. It has mentioned to facilitate the approval for innovative drugs and partially granted generics the priority in the approval procedure.

3. 3. Biosimilar regulatory standards

However, a biosimilar is a medicinal product of biological origin, usually with a complex structure, and large proteins (such as antibodies) that are subject of heterogeneity.

European Union has established a guideline for biosimilar to facilitated its Member States to put biosimilar under national legislation. In the European Union (EU), decisions on pharmacy substitution of biosimilars are made at the Member State level. France is the first European country to explicitly permit biosimilar substitution. In France, as of 1st January, 2014 new provisions have been in force concerning biosimilars. Article 47 of the Law of 23 December, 2013, concerning the budget of the Social Security, has adapted provisions which allow the substitution of originators by generics to extend to biosimilar products.

On the other side of the continent, in China, there is little on regulating biosimilars. Chinese has passed the law on managing drugs in 2001, and the latest follow on rulings is 7 years ago, and there is no biosimilar guidelines like in Europe. The only measures for biosimilars, are to use the new drug procedures in order to be granted the marketing approval. That means for a biosimilar from R&D to manufacturing it will be awfully long, and money consuming. In the past 3 years there are over 1 000 new bio - drugs applications but only 70 has made to the end[1].

3. 4. Conclusion

France will now be allowed to substitute a biosimilar for the prescribed (reference) biological under certain conditions, including only when initiating a course of treatment and that the prescribing physician has not marked the prescription as "non - substitute". Legislation allowing substitution of biosimilars has been introduced as part of a new law concerning the social security budget (Article 47 of the Law of 23 December, 2013), which came into effect on 1 January, 2014. Provisions in the law that allow the substitution of originators by generics have been adapted to extending to biosimilars.

---

[1] Report on biosimilar related R&D guidelines and regulations, Sino Medicine, 12 July, 2014, http: //zhongyi. sina. com/gengnianqi/zyjd/104. shtml.

On the other side of the continent, in China, there is little on regulating biosimilars. On 22 February last year, the SFDA in China has released "suggestions on encouraging innovative drugs", it is said to improve the administration approval on innovative drugs and the valuable therapeutic generics. It has mentioned to facilitate the approval for innovative drugs and partially granted generics the priority in approval procedure. We have to be awared that even though the suggestions have no enforceable bindings as legislation, but in the practice it is used as law applicable only if there is another higher authority establish another piece of administration doctrine that is contrary to the prior.

To draw the conclusion on biosimilars in China, it is put into administration documents as generics, and with no prejudice to the generics in the EU consensus. It is controlled regulated by State Food and Drug Administration (SFDA) for all related matters. There is no specific legislation governing on either generic or biosimilar.

We can see in this paper that the biosimilar is certainly not a generic, so there is no rational for calling a biosimilar as bio – generic. At the regulatory stage, a biological drug is more complex and perhaps less clear than in the stage of small molecule drugs. The underlying nature of science for biologics is also in some ways more complex than for small molecule drugs. It is the intersection of the legal framework and the science that makes generic drugs the "same" as their reference products and biosimilars "similar" to their pioneers. The reason to make the simple complicated is because, biosimilar itself needs to be listed under an existing category in order for legislator, practitioners, manufactures and consumers to be clearly understood and keep in mind what drugs they are dealing with.

IV. Conclusion

Is biosimilar the same as generic, a copy to the brand name drug or it is simply just another drug with bio – similarity to the brand name drug? This paper is trying to discuss this specific question in exploring exclusive rights in pharmaceutical industry.

Patent issue is important and essential to the pharmaceutical industry. It is due to the nature of highly innovative industry, and huge amount of investment on each drug from an invention to a product sell on the market.

Exclusive rights in pharmaceutical industry is important, it has the common feature of exclusive rights in tangible and intangible goods, it is a bundle of rights granted to perform or to prevent others to perform, but with specific standard in practice. It is the small and tiny specifics that make exclusive rights in drugs a debatable question at UN level. The rationality for intellectual property and for public health is controversial, both sides are trying to reach their goals either different strategies in patent drugs and their prices, or setting restrictions to the drug circulation. However, the solution given by the international treaties is not efficient in practice, and it leads to many problems between states.

The industry itself has tried to find some new solutions for their net profit continuously increasing. These strategies seem to be crucial at glance, but in reality it worked. Hence, generics came into sights, later on as biotechnology developed during last decades biosimilar became another substitute to patent drug. Europe as one of the first region in the world to guide the application of biosimilar, has set many examples on how to handle this issue. As it is seen in this paper, the guideline did not set up a standard, they use a case – by – case method. Chinese State Food and Drug Administration (SFDA) has considered biosimilar as a new drug, and needs to apply under the new drug process. Biosimilar has no advantage of facilitate the companies investment compared with brand name drug itself. France has recognized biosimilars as substitutes to brand name drugs, but it has debate over the issue.

One of the results of this game between two parties is the result of copying drugs from patent or blockbuster drugs. Technology has improved and grew rapidly, it has gone from small molecule drugs to biotechnology oriented drugs. Thus, the issue of copying also changes from generics to biosimilars.

Thus, the international organizations should take on a leading role, coordinating global health efforts with long - term vision, and a concrete but flexible plan for implementation. And for the pharmaceutical companies, the question is whether they choose to stick on their existing patent drug, and apply to a second therapeutic application[1] to have another patent on the same drug. Or should they take the path of the second generation on patent drug, to invest on generics and biosimilars?

A good way to settle this at international level is to take second generation on patent drugs as a topic to discuss like any other topics in to forum or ministerial meetings.

Firstly, the international entity should set an R&D agenda for the world, prioritizing diseases based on which ones constitute major global health concerns and the burden imposed by diseases and adjusting priorities as necessary.

Secondly, the differential pricing between developed and developing countries to be established as routine in case of essential drugs, and what are the alternatives.

Thirdly, the most important to the question is positioning on the role of biosimilars. Whether it shall be given the exclusive rights like other copies of patent drug, or it shall be granted just as another new patent drug, or should their be a third category.

---

[1] Emphasis was placed on the wording of Article 54 (5) EPC, which refers to the maintainability of a substance for any specific use prohibited by Article 53 (c). To impose the requirement that the "specific use" related to the treatment of a different disease would be to arbitrarily introduce a distinction, which the legislation does not make.

LEGAL ISSUES OF INTERNATIONAL MEDICAL TOURISM TO CHINA

Dean M. Harris, J. D. [1]

Abstract:

The People's Republic of China has taken steps to increase its role as a destination for medical tourism. However, China has also taken steps to prevent "transplant tourism," in which relatively wealthy foreigners could obtain organ transplants from donors in China. On the surface, these two efforts may appear to be inconsistent, because one would increase medical tourism to China in general while the other would prevent one type of medical tourism. However, these two efforts are actually consistent and complementary. By taking steps against transplant tourism, which has been strongly criticized by international observers on ethical grounds, China has increased respect for its health care system. In that way, China has increased the likelihood that residents of other countries will look to China as a potential destination for medical tourism in the future.

This paper begins by analyzing the causes and characteristics of medical tourism in general. Next, the paper analyzes the legal issues of medical tourism including legal recourse for medical malpractice, the legal framework to promote and regulate medical tourism, laws that would protect residents of a destination country, and laws that would protect residents of a patient's home country. Then, the paper applies these facts and legal issues to analyze China as a destination for medical tourism from both practical and legal perspectives.

[1] Department of Health Policy and Management, Gillings School of Global Public Health, University of North Carolina at Chapel Hill, Chapel Hill, North Carolina, USA, 27599 -7411 Dean_ Harris@ unc. edu.

1. International Medical Tourism: Causes and Characteristics

The term "medical tourism" is used to describe the practice of traveling to a country other than the patient's country of residence for the purpose of obtaining medical treatment. [1] The term covers a broad range of different practices as well as different reasons to travel for medical care. Sometimes, individuals travel to other countries for the purpose of obtaining care of higher quality. [2] In other situations, individuals who reside in industrialized countries travel to less developed countries to obtain medical care for various reasons.

One reason for traveling to a less developed country is the significantly lower cost of care. [3] In a 2014 article, Ye and colleagues noted that the average cost for a heart bypass procedure in U. S. dollars varied from \$ 130 000 in the United States to \$ 18 500 in Singapore, \$ 11 000 in Thailand, \$ 10 500 in Shanghai, and \$ 10 000 in India. [4]

However, reasons other than cost exist for traveling to another country for medical treatment. Another important reason is to obtain a treatment that is not available in a patient's home country. [5] For example, some countries allow a patient more flexibility than the patient's home country to choose assisted reproductive technology

---

[1] U. S. Centers for Disease Control and Prevention (CDC), "Medical Tourism – Getting Medical Care in Another Country", http: //www. cdc. gov/features/medicaltourism/

(accessed August 9, 2014). *See generally*, Harris, D., Ethics in Health Services and Policy: A Global Approach, (Jossey – Bass/John Wiley & Sons, Inc., San Francisco, CA, 2011), 206 – 211, < http: // www. wiley. com/WileyCDA/WileyTitle/productCd – 0470531061. html > (accessed August 9, 2014); Cortez, N. "Patients Without Borders: The Emerging Global Market for Patients and the Evolution of Modern Health Care". *Indiana Law Journal*, (2008), 83: pp. 71 – 132;

Bookman, M., and Bookman, K. Medical Tourism in Developing Countries (New York: Palgrave Macmillan, 2007).

[2] See Harris (2011), supra note1, at 206.

[3] *See generally*, Fried, B., and Harris, D., "Managing Healthcare Services in the Global Marketplace". Frontiers of Health Services Management (2007), 24 (2), p. 6.

[4] Ye, B., et al, "An Assessment of Medical Tourism Development Potential in Mainland China", *Public Administration and Policy: An Asia – Pacific Journal* (Spring 2014), 17 (1): pp. 32 – 40, at 35. http: //journal. hkpaa. org. hk/index. php/paap – journal/spring – 2014 (accessed August 9, 2014).

[5] Cortez (2008), *supra* note 1, pp. 77 – 78.

(ART) or abortion. [1] As stated by an organization in the United Kingdom that supports choice in abortion, "As abortion is restricted in Ireland, if you decide to have an abortion you must travel to the UK and pay for the service." [2] In addition, a patient might travel to another country to obtain a drug that has not been approved for use in the patient's home country. [3] Other reasons that patients might choose to travel for medical care is to have greater privacy in receiving treatment such as cosmetic surgery or to combine medical treatment with a vacation in a foreign country. For example, one medical tourism company offers "Privacy in Paradise" as well as "Surgery and Recuperation with time to heal away from public scrutiny." [4]

Several countries have become popular destinations for medical tourism, including Thailand, India, and Singapore. [5] In the Western Hemisphere, destination countries for medical tourism include Brazil, Mexico, Argentina, Costa Rica, and

---

[1] Id. p. 77.

[2] Marie Stopes International, "Travelling from Ireland", http://www.mariestopes.org.uk/Womens_services/Abortion/Abortion_for_women_from_Ireland_and_other_countries/Travelling_from_Ireland.aspx (accessed August 9, 2014). See also Daly, B, " 'Braxton Hick's' or the Birth of a New Era? Tracing the Development of Ireland's Abortion Laws in Respect of European Court of Human Rights Jurisprudence", *European Journal of Health Law* (2011), 18 (4): pp. 375 - 395, p. 381 and p. 389 (describing court cases in which individuals in Ireland had travelled to England for abortion).

[3] Cortez (2008), supra note 1, p. 77 - 78. For example, the U.S. Food and Drug Administration (FDA) recognizes that "drugs from other countries that are available for purchase by individuals often have not been approved by FDA for use and sale in the United States". U.S. Food and Drug Administration, "Is it legal for me to personally import drugs?", http://www.fda.gov/AboutFDA/Transparency/Basics/ucm194904.htm (accessed August 9, 2014). Although importation of drugs into the United States by individuals is usually unlawful, the FDA usually does not object to importation of small amounts of unapproved drugs by individuals under some circumstances. Id.

[4] Surgeon & Safari CC, "Welcome", http://www.surgeon-and-safari.co.za/index.html (accessed August 9, 2014). The website of that company also lists various types of holidays including safaris, day trips, and a shopping excursion. Id.

[5] Ye, et al, supra note 4, p. 32.

Cuba. [1] "The most common categories of procedures that people pursue during medical tourism trips are cosmetic surgery, dentistry, cardiology (cardiac surgery), and orthopedic surgery." [2]

Some hospitals in destination countries have taken steps that could gain the confidence of medical tourists and payers by achieving international accreditation through an organization, such as Joint Commission International (JCI). [3] The U. S. Centers for Disease Control and Prevention (CDC) has emphasized the importance of accreditation for prospective medical tourists.

"Travel health providers should advise prospective medical tourists to determine if health care facilities they are considering are accredited by the Joint Commission International (JCI). JCI is the international division of the Joint Commission Resources, a US – based, not – for – profit affiliate of the Joint Commission that certifies health care facilities in the United States. On March 2012, JCI has accredited 368 international hospitals in 46 countries …. As more facilities are accredited, more providers will likely offer incentives for their patients to travel overseas for care." [4]

Estimates of the number of medical tourists vary greatly. According to the United States CDC, "little reliable epidemiologic data on medical tourism exist. Studies using different definitions and methods have estimated there are 60,000 – 750,000

[1] U. S. Centers for Disease Control and Prevention (CDC), "Yellow Book 2014: Medical Tourism", http: //wwwnc. cdc. gov/travel/yellowbook/2014/chapter – 2 – the – pre – travel – consultation/medical – tourism (accessed August 9, 2014). See also Pitts, B and N Battiste, "As More Americans Have Surgeries Overseas, US Companies Consider 'Medical Tourism' a Health Care Option", ABC News (September 30, 2013), http: //abcnews. go. com/Health/americans – surgeries – overseas – us – companies – medical – tourism – health/story? id = 20423011 (accessed August 9, 2014).

[2] U. S. Centers for Disease Control and Prevention (CDC), "Yellow Book 2014: Medical Tourism", supra note 11.

[3] See Joint Commission International (JCI), "JCI – Accredited Organizations", < http: //www. jointcommissioninternational. org/about – jci/jci – accredited – organizations/ > (accessed August 9, 2014).

[4] U. S. Centers for Disease Control and Prevention (CDC), "Yellow Book 2014: Medical Tourism", supra note 11.

medical tourists annually from around the world." [1] In that regard, an article in The Economist in 2014 explained the problem of estimation and the poor quality of available data on medical tourism.

"In 2008 Deloitte predicted an 'explosive' boom in medical tourism, saying that the number of Americans going abroad for health care would grow more than tenfold by 2012.

It did not happen. Poor data were part of the problem: whereas Deloitte counted 750, 000 American medical tourists in 2007, McKinsey, another consultancy, found at most 10, 000 a year later. It is generally agreed that the number of medical tourists has grown since then—Thailand's Bumrungrad hospital, which is popular with foreign patients, reports 'steady growth'. But the data are still fuzzy. Patients Beyond Borders estimates that as many as 12m people globally now travel for care, perhaps 1m of them Americans. Industry insiders admit that growth has not matched the initial heady expectations." [2]

One possible reason for the poor quality of data on medical tourism may be the difficulty of determining whether a non - resident patient who received treatment in a foreign country actually traveled to that country for the purpose of receiving treatment. Hospitals are likely to maintain data on the residence of each patient, and hospital managers probably pay close attention to data on their patient origin and geographic service area. However, for patients who are citizens of a different country, some hospitals might not distinguish among: (a) patients who traveled specifically to obtain treatment; (b) expatriate workers or retirees who resided at that time in the country in which the hospital is located; and, (c) tourists, visitors, or business travelers who became sick or injured while in that foreign country for purposes other

---

[1] Id.

[2] The Economist, "Medical tourism: Médecine avec frontiers: Why health care has failed to globalise", (February 15, 2014), <http://www.economist.com/node/21596563/print> (accessed August 9, 2014).

than obtaining treatment. Thus, the data on medical tourism are questionable, and it is not surprising that people disagree.

2. Legal issues of International Medical Tourism

Most discussions about legal issues of international medical tourism focus on the problems in obtaining legal relief in the event of medical malpractice. That is an important issue, but it is not the only legal issue. Other important issues include the legal framework to promote and regulate medical tourism, laws that would protect residents of the destination country, and laws that would protect residents of the patient's home country. Each issue is addressed below.

A. Legal recourse for medical malpractice

If a medical tourist was injured by negligence in a foreign country, the medical tourist would probably not be able to pursue a claim against the foreign health care provider in the courts of the medical tourist's home county. [1] The medical tourist would probably have to pursue any claim in the courts of the foreign country. Legal rights and remedies in foreign courts might be very limited, and are likely to be very different from the rights and remedies in the medical tourist's home county. [2] As Cortez has explained, "I conclude that U. S. medical tourists will struggle to obtain adequate compensation, either here or abroad. Patients looking to sue in U. S. courts for medical malpractice abroad will face difficulties locating a proper defendant, venue, and theory of liability. Patients suing overseas will also face obstacles recovering adequate, timely compensation in legal systems that use unfamiliar procedures, communicate in foreign languages, limit the remedies available, and impose more onerous burdens of proof." [3]

Under these circumstances, some government agencies and medical professional

---

[1] Cortez, N., "Recalibrating the Legal Risks of Cross – Border Health Care", *Yale Journal of Health Policy, Law, and Ethics* (Winter 2010), 10: 1 – 89 (analyzing the legal barriers to pursing claims in U. S. courts against foreign health care providers, intermediaries, insurance companies, and employers).

[2] Id. (analyzing the legal rights and remedies in Thailand, Singapore, India, and Mexico).

[3] Id. p. 5.

associations have warned potential medical tourists about the type of legal recourse that would be available to them in other countries. [1] For example, the American College of Surgeons adopted a position statement on medical tourism which includes the warning that "Patients should be aware that many of the means for legal recourse available to citizens in the U. S. are not universally accessible in other countries." [2] It is ironic that this professional association would warn potential medical tourists that they might lose the advantages of the U. S. legal system in cases of malpractice. This association, like many others, advocates for changes to the U. S. liability system to fix what it describes as a nationwide "medical liability crisis." [3]

Moreover, most patients who are injured by negligent treatment in the U. S. do not pursue a claim for medical malpractice, even though they have the right to do so. [4] Although medical tourists from the U. S. have to give up some of their legal rights and remedies, those rights and remedies are not used by most patients injured by negligent treatment in the United States.

B. The legal framework to promote and regulate medical tourism

A government can adopt various types of laws and regulations to promote medical tourism to its country. This legal framework could include laws that create special zones for medical tourism. [5] Within those special zones, laws or regulations could provide tax advantages, faster regulatory approvals, long – term visas for foreign health care professionals, and greater flexibility for investment of foreign capital and

[1] See, e. g., U. S. Centers for Disease Control and Prevention (CDC), "Medical Tourism – Getting Medical Care in Another Country", supra note 1 ("Determine what legal actions you can take if anything goes wrong with the procedure"); Cortez (2010), supra note 17, p. 18 and n. 97 (discussing a warning from the National Health Service of the United Kingdom).

[2] American College of Surgeons, "Statement on Medical and Surgical Tourism", 2009, https://www.facs.org/about – acs/statements/65 – surgical – tourism (accessed August 10, 2014).

[3] American College of Surgeons, "The Medical Liability Crisis" (2011), https://www.facs.org/advocacy/federal/liability/the – medical – liability – crisis (accessed August 10, 2014).

[4] Harris, D., Contemporary Issues in Healthcare Law & Ethics (Fourth Edition), Chicago, Health Administration Press, 2014, p. 249.

[5] See Ye, et al, supra note 4, p. 33.

establishment of foreign medical facilities. [1] In addition, laws and regulations could simplify the requirements and procedures to issue visas for medical tourists.

To protect medical tourists and promote medical tourism, a legal framework should include an effective system of licensing for all health care professionals and facilities that serve medical tourists. [2] Some experts have also recommended that licensing should be required for intermediaries that facilitate the process of medical tourism. [3] Another problem in medical tourism has been the lack of adequate medical records from foreign health care providers for follow – up care in the patient's home country. [4] This situation could be improved if governments of destination countries would adopt requirements to standardize medical documentation and reaffirm the patient's right to obtain a copy of the medical record. [5]

Finally, a legal framework for medical tourism should include laws and regulations to improve the systems of quality assurance and resolution of medical disputes. [6] As Ye and colleagues have stated, "Regulating the medical tourism market is crucial for sustainable development. The government should formulate entry requirements, quality control system, and other policies to regulate the market and build up customer confidence. For example, the government may consider emulate India's practice to establish a star – rating system based on each hospital's service quality." [7]

---

[1] Id.

[2] Id. p. 37.

[3] Crooks, V, et al, "Ethical and legal implications of the risks of medical tourism for patients: a qualitative study of Canadian health and safety representatives' perspectives", *BMJ Open* (2013), 3: e002302. doi: 10.1136/bmjopen – 2012 – 002302. http://bmjopen.bmj.com/content/3/2/e002302.full.pdf + html (accessed August 10, 2014), p. 6.

[4] Id. p. 4. "Participants learnt from clinicians and other front – line providers that patients returning from medical care abroad often bring back incomplete medical records or no documentation of the care they received. Although some clinicians have requested results and records from international medical providers, destination facilities were often uncooperative." Id.

[5] Id. p. 4, 6.

[6] Ye, et al, supra note 4, pp. 37 – 38.

[7] Id. p. 38.

C. Laws that would protect local residents of a destination country

Medical tourism could affect residents of a destination country by reducing access to care for local patients. Medical tourism can create financial incentives for physicians to move from public facilities that treat local residents to private hospitals that treat many medical tourists and, therefore, can afford to pay higher salaries. [1] Medical tourism could also use a disproportionate share of other limited resources such as hospital beds, equipment, and drugs. In addition, the practice of "transplant tourism" can encourage vulnerable residents of destination countries to sell their organs to foreigners, or can make it possible for foreigners to obtain some organs that otherwise might have been available to local patients who need transplants. [2] Transplant tourism has been widely criticized by ethicists and international organizations. [3] "Increasing transplant availability is a worthy goal, but transplant tourism is not an ethical approach." [4]

In theory, medical tourism could help a destination country and its residents by promoting economic growth and generating revenue that might help to subsidize health services for local residents. [5] To accomplish these goals, governments of destination countries would need to adopt and enforce a legal framework to regulate medical tourism and require by law that some revenues be used to subsidize local services. Laws and regulations could require health care providers that serve medical tourists to provide a specific volume of service to local patients or pay taxes designat-

---

〔1〕 Harris (2011), supra note 1, p. 208 (referring to this phenomenon as "internal brain drain").

〔2〕 Id. p. 210.

〔3〕 Id. See also Budiani - Saberi, D., and Delmonico, F., "Organ Trafficking and Transplant Tourism: A Commentary on the Global Realities", *American Journal of Transplantation* (2008), 8: pp. 925 - 929; Bramstedt, K., and Xu, J., "Checklist: Passport, Plane Ticket, Organ Transplant", *American Journal of Transplantation* (2007), 7: pp. 1698 - 1701.

〔4〕 Bramstedt, K., and Xu, J, supra note 34, p. 1700.

〔5〕 Harris (2011), supra note 1, p. 208.

ed for the support of public health care facilities. [1] Another alternative is to impose a tax on the medical tourist or intermediary, by adding a surcharge to the fee for a visa and designating that surcharge for use of public facilities in the destination country. Finally, the legal framework should prohibit unethical practices such as transplant tourism.

D. Laws that would protect residents of a patient's home country

A medical tourist's home country and residents of that home country could be harmed by medical tourism in at least two important ways. Firstly, a medical tourist could return to the home country with a communicable disease or antibiotic – resistant organism that presents a threat to other individuals and the public health. [2] Secondly, a medical tourist might impose the costs of follow – up care, including the cost of treating complications, on the health system and taxpayers of the medical tourist's home country. [3]

With regard to the problem of communicable disease and antibiotic – resistant organisms, the government of the home country has some potential legal responses. However, a government cannot simply prohibit its residents from leaving their home country and traveling to another country for medical care. According to the United Nations' Universal Declaration of Human Rights, "Everyone has the right to leave any country, including his own, and to return to his country." [4] Nevertheless, when medical tourists return to their own country, the government of the home coun-

---

[1] Id. p. 208 – 09; Mattoo, A. and R Rathindran, "How Health Insurance Inhibits Trade in Health Care", *Health Affairs* (2006), 25 (2): pp. 358 – 368, p. 367; Chinai, R. and R Goswami, "Medical Visas Mark Growth of Indian Medical Tourism", *Bulletin of the World Health Organization* (2007), 85 (3): pp. 164 – 165, p. 165.

[2] Crooks, et al, supra note 27, p. 4. See also Bramstedt and Xu, supra note 34, p. 1700.

[3] Crooks, et al, supra note 27, pp. 5 – 6. In economic terms, the industry of medical tourism has externalized the inevitable costs of medical complications and follow – up care, and has shifted those costs to the home country and its taxpayers.

[4] United Nations, "Universal Declaration of Human Rights", Article 13 (2), http://www.un.org/en/documents/udhr/index.shtml (accessed August 11, 2014).

try can impose requirements to protect the public health such as isolation, quarantine, and surveillance of individuals who have been exposed to a communicable disease.

It may be more difficult to develop a legal response to the problem of imposing follow – up costs on a home country and its residents. Ordinarily, governments of developed countries do not require individuals to bear all of the costs of their risky behavior, even if that behavior imposes burdens on society. For example, governments can try to reduce smoking and obesity, which can increase health care costs for society. However, governments and health care providers generally do not impose higher fees for health care services on individuals who smoke or refuse to exercise.[1] Home countries are not likely to recover all of the additional costs of follow – up care and complications from individual medical tourists.

Instead, governments of home countries could try to hold the industry of medical tourism liable for the additional costs that home country governments will incur. For example, state governments in the U. S. reached a settlement with tobacco companies to impose restrictions on those companies and provide for payments to the state governments. [2] By analogy, governments whose residents travel to other countries for medical treatment might be able to require payments from medical tourism companies. Those payments could help pay the additional costs that medical tourists impose on their societies after they return home.

3. China as a destination for International Medical Tourism

In 2007, an American newspaper reported that hundreds of foreign residents had gone to China for stem cell injections that were not allowed in the United

[1] See generally Pearson, S. and S. Lieber, "Financial Penalties For The Unhealthy? Ethical Guidelines For Holding Employees Responsible For Their Health", *Health Affairs* (2009), 28 (3): pp. 845 – 852 (discussing the related ethical issue of personal responsibility and financial incentives in employee health plans).

[2] National Association of Attorneys General, "Tobacco", http://www.naag.org/tobacco.php (accessed August 11, 2014).

States.[1] As of 2014, however, experts recognized that China had not yet developed a mature industry of medical tourism and had not yet established itself as a desirable destination for that purpose.[2] According to People's Daily Online on April 29, 2014, "The medical tourism industry in China is still at the beginning."[3] Ye and colleagues recognized the lack of progress, but they also recognized China's potential to develop medical tourism and the recent steps taken to achieve that goal.

China has the potential to become the regional medical tourism destination. Firstly, traditional Chinese medicines and therapies (e. g., acupuncture, cupping) have a long history and have become a popular form of alternative medicine/treatment in China and many other countries (e. g., U. S.). Secondly, China also has abundant natural and cultural resources for tourism which can enrich tourist experiences during their journey. Thirdly, the medical tourism expenditure in China is rather low as compared with developed countries. Despite of the aforementioned advantages, China has not yet fully leveraged its resources and its medical tourism development is in infant stage.

By far, there have been some cities (e. g., Sanya, Shanghai) that launched medical tourism by offering particular medical service (e. g., traditional Chinese medicine) to overseas patients and some provinces and cities (e. g., Hainan province) have implemented strategic plans to develop themselves into medical tourism hub. For example, Hainan province plans to establish "Boao Lecheng" medical tourism centre, covering services such as elderly care, cosmetic and plastic surgery, traditional Chinese medicine, chronic diseases rehabilitation and sub-health sanato-

---

[1] Zamiska, N, "Seeking Hope In Chinese Stem Cells", *Wall Street Journal* (July 2, 2007), p. B1.

[2] Ye, et al, supra note 4, p. 32, 37. "The medical tourism development in China is in its infancy stage... Therefore, visiting China to obtain medical care is still not a preferred option for many medical tourists." Id., p. 37.

[3] People's Daily Online, "Medical Tourism Starts in China", <http://english.peopledaily.com.cn/98649/8612990.html> (accessed August 14, 2014).

rium. [1]

In addition to these steps to promote medical tourism in general, China has taken steps to prevent "transplant tourism". In 2007, China's State Council adopted Regulations on Human Organ Transplantation that prohibit "trade in human organs". [2] In addition, that regulation limits organ transplants from live donors to relatives of the donor or individuals who have a close relationship with the donor. [3] According to Huang and colleagues, the 2007 regulation prohibits transplant tourism, but nevertheless that practice has continued. [4] Similarly, Shi and Chen wrote in 2011 about illegal organ trading in China, despite the issuance of principles by the Ministry of Health in December of 2009 against trade in organs from live donors. [5]

On the surface, China's efforts to promote medical tourism and prevent transplant tourism may appear to be inconsistent, because one effort would increase medical tourism in general while the other would prevent one type of medical tourism. However, those two efforts are actually consistent and complementary. By taking steps to reduce transplant tourism, which has been strongly criticized by international observers on ethical grounds, China has increased respect for its health care system. In that way, China has increased the likelihood that residents of other countries will look to China as a potential destination for medical tourism in the future.

Has China adopted the legal and regulatory structure to promote medical tourism and protect the interests of all parties? If not, what types of laws and regulations

---

[1] Ye, et al, supra note 4 p. 33 (reference omitted).

[2] National Health and Family Planning Commission of the PRC, "Regulations on Human Organ Transplantation", Article 3, http://www.chinadaily.com.cn/m/chinahealth/2014-06/05/content_17573845.htm (accessed August 15, 2014).

[3] Id. at Article 10.

[4] Huang, J, et al, "A Pilot Programme of Organ Donation after Cardiac Death in China", Lancet, (March 3, 2012), 379: pp. 862-865, p. 862.

[5] Shi, B and L Chen, "Regulation of Organ Transplantation in China: Difficult Exploration and Slow Advance", *Journal of the American Medical Association* (2011), 306 (4): pp. 434-435, p. 434.

should China consider?

With regard to laws about medical malpractice and resolution of doctor – patient disputes, Chinese authorities have made some significant reforms, but additional reforms are necessary. The 2002 Regulation on the Handling of Medical Accidents had made some improvements over the previous system. However, the 2002 regulation left several issues undecided, and it did not establish a system of dispute resolution that was trusted by the people. [1] More recently, China has enacted a new Tort Liability Law that became effective on July 21, 2010 and explicitly addressed medical liability. [2] Meanwhile, Chinese authorities have been working to implement mediation as a method of alternative dispute resolution (ADR). [3] Nevertheless, China still has a problem of protests by patients and their families as well as incidents of violence against physicians. [4] As noted by Ye and colleagues, violence against physicians and deficiencies in the system for handling medical disputes are weaknesses that need to be addressed, in order to make China a more attractive destination for medical tourism. [5]

Apparently, China does not have a unified, comprehensive law to promote and regulate medical tourism. China has adopted some rules or policies to promote medical tourism, by methods such as tax incentives and flexibility for foreign investment,

---

[1] Harris, D. and C. C. Wu, "Medical Malpractice in the People's Republic of China: The 2002 Regulation on the Handling of Medical Accidents", *Journal of Law, Medicine & Ethics* (2005) 33 (3): pp. 456 – 477.

[2] "Tort Liability Law of the People's Republic of China", *Journal of European Tort Law* (2010), 1 (3): pp. 362 – 375 (translation by Yan Zhu and Helmut Koziol). See generally, Wang, Z. and K. Oliphant, "Yangge Dance: The Rhythm of Liability for Medical Malpractice in the People's Republic of China", *Chicago – Kent Law Review* (2012) 87: pp. 21 – 52.

[3] See, e. g., Wang, H, "Shanghai to launch mediation system for medical disputes", chinadaily. com. cn, (Updated February 21, 2014), http://www.chinadaily.com.cn/cndy/2014 – 02/21/content_17295642.htm (accessed August 14, 2014).

[4] Liebman, Benjamin L., "Malpractice Mobs: Medical Dispute Resolution In China", *Columbia Law Review* (2013), 113: pp. 181 – 264.

[5] Ye, et al, supra note 4, pp. 37 – 38.

but those rules or policies only apply to special zones such as the Boao Lecheng International Medical Travel Zone in Hainan province. [1] In addition, China could consider a system of licensure for companies that arrange medical tourism, [2] and could improve the enforcement of existing laws on licensure for health care professionals. [3] Finally, China could simplify its visa process for medical tourists by creating a special category of "medical tourism visitor" with expedited procedures for issuance.

With regard to laws to protect local residents from the effects of medical tourism, China has already taken important steps to prevent transplant tourism, as discussed above. [4] Additional steps could include regulations for taxation of medical tourists, providers, and companies that arrange medical tourism, in order to provide more financial support for local public services.

As a destination country, China could adopt some regulations that would promote medical tourism in China and indirectly protect a home country and its residents. For example, follow - up costs in a medical tourist's home country might be reduced if medical tourists could return home with detailed records of their treatment in a destination country. Therefore, China could help home countries and promote China as a destination by adopting legal requirements to standardize medical documentation and reaffirm the patient's right to obtain a copy of the medical record. [5]

---

[1] See, People's Daily Online, "Hainan plans medical tourism zone", (April 7, 2013) http://english.peopledaily.com.cn/205040/8196686.html (accessed August 15, 2014). See also Ye, et al, supra note. According to Ye and colleagues, the limitation of these policies to special zones is a weakness in China's effort to promote medical tourism. Id. pp. 36 - 37.

[2] *See* note 27 and accompanying text *supra*.

[3] See generally, Lim, M. K., et al, "Public Perceptions of Private Health Care in Socialist China". *Health Affairs* (2004), 23 (6): pp. 222 - 234. In fact, Ye and colleagues noted that some medical tourists have received treatments of traditional Chinese medicine (TCM) from non - licensed providers or facilities that were not medical institutions. Ye, et al, supra note 4, p. 37.

[4] See, notes, pp. 47 - 50 and accompanying text supra.

[5] See, notes, pp. 28 - 29 and accompanying text supra.

Conclusion

China has the potential to develop an industry of medical tourism and promote the country as a destination for medical tourists. China has taken some important steps toward these goals, including establishing policies for special zones and adopting regulations against transplant tourism. Future efforts should include creating a comprehensive legal framework to promote and regulate medical tourism for the protection of all parties.

PROTECTION OF INTELLECTUAL PROPERTY AND TRADITIONAL CHINESE MEDICINE: PATENT SYSTEM IN CHINA

ZHUANG Chuanjuan[1]

Abstract:

Traditional Chinese Medicine (TCM) is a treasure of Chinese culture, which needs to be protected. The Chinese intellectual property legal system is relatively complete, however it is hard to protect TCM, especially due to its ancient and traditional nature. That is because the main purpose of patent law is more to protect innovation and novelty rather than tradition.

This article is dedicated to the study of the law applying to TCM in China, and also observes and analyzes the potential and benefits that patent law could bring in protecting TCM. The study concedes that the theories of TCM are not protected legally by patent. It also notes that the potential to protect the technical aspects of TCM remains very limited, because existing restrictive conditions in the law itself are relatively unfavorable.

To conduct the study, the author initially analyzes how a patent could protect TCM, then examines the difficulties in using the patent protection for TCM.

Keywords: TCM, Intellectual Property, Patent, Novelty, Creativity, Practicality

Traditional Chinese Medicine (TCM), as a treasure in traditional Chinese culture, has made tremendous contribution to the field of human reproduction, disease prevention, health care, etc. And it is one of the most complete traditional medicine

[1] Ph. D student at Toulouse 1 Capitole University, France.

systems in the world up to now[1]. With the development of research on human diseases in the field of life sciences, TCM has attracted wide attention from the international medical profession.

Current intellectual property system includes patents, trademarks, copyright, unfair competition, etc. The "People's Republic of China Patent Law" was implemented on April 1st, 1985 and adopted at the fourth meeting of the Standing Committee of the National People's Congress Sixth People's Republic of China on March 12th, 1984. From then on, the inventions in the field of medicine were administrated by the method of patent protection. However, the law stipulates that: "Chemicals and substances obtained by chemical methods are not patentable." In order to further strengthen the protection of medicine, the amended "People's Republic of China Patent Law" was implemented on January 1st, 1993, and expanded the scope of protection of the patent, beginning to grant patent protection for medicine[2].

Since TCM is completely different from western medicine both in medical technology and culture, it is not easy for most TCM knowledge to obtain patent protection as pharmaceuticals.

Section one: Characteristics and scope of the TM (Traditional Medicine) knowledge

As everyone knows, TCM is a huge complex knowledge system. It is rich in content, forms of existence, means of expression, related resources, and so on. TM theory is valuable because it comprises the concept of holism and treatment based on syndrome differentiation. The theoretical system is macro and integrity, and also pays attention to individualized. TM also has inadequacies, mainly regarding the following two ways:

---

[1] Qingming Zhao, Guangfu Xiang, "TCM Concept undert the Legal Perspective", *People's Forum*, 2010 (13), p. 92.

[2] Xiurui Bian, "Thinking about the intellectual property protection of TM (Traditional Medicine)", *Chinese Pharmaceutical Affairs*, 2004 (4), p. 232.

First, TM theory neglects microscopic study. For a long time, TM was based on human health and disease by understanding and transforming objects through practice and theoretical thinking.

Due to these historical conditions, most qualitative content does not have enough quantitative analysis to be support, and at the macro - level the understanding lacks the basis of the micro - level. The general theory lacks precise exposition and has no empirical analysis or an understanding of the details of the internal structure of the human body. TM only emphasizes macroscopic integrity, while the elements, the changes and the links between the elements, the media of the elements, do not do possess enough supporting research.

Second, TM lacks quantitative analysis and standardization. TM emphasizes individualized treatment, which is difficult for research design and for choosing the control group. Treatment based on syndrome differentiation theory emphasizes a main syndrome, and then focuses on the smaller syndrome, while paying attention to syndrome differentiation. Only by correct judgment regarding disease syndromes, is the physician able to dispense a reasonable prescription. The patient's complaints and symptoms always have considerable arbitrariness and individual differences, and are also affected by the ability of the patient to express themselves. As information is received by the inspecting, smelling, inquiring and pulse - taking method, it has a large degree of subjectivity. Judgments also might maintain subjective one - sidedness. As dialectic was based on traditional diagnostic technology and clinical experience from different doctors, TM was difficult to standardize.

Diagnostics of TCM is one of the important research of basic theory of TCM. Its lack of quantify is an important factor which restricting the development of TCM. Chinese materia medica has a completely different mechanism of action to Western medicine, the healing characteristics was especially strengths embodies on the compound Chinese Medicine. Pharmaceutical patent protection dominated by chemicals in the developed countries, and natural pharmaceutical composition related patent examination has no determining criteria, and no recognized international standards.

There's a variety of different perspective about the scope of TM including[1]. In this article, the author intends to elaborate the TM knowledge as two parts, theoretical knowledge and technical knowledge.

1. Theoretical knowledge

The theoretical knowledge of TCM is the headstone for the survival and development of Chinese medicine. It is also a science in relation to the normal body along with different laws of states of life and their variations. Theoretical knowledge includes the knowledge of life, diseases, health care and drug theory, principle of TCM action, law of compound compatibility, etc. Obviously, TCM theoretical knowledge is not protected by patent law. Due to the theory of patenting, scientific theory can not be patented. Once granted a patent, the development of scientific technology would be hindered.

2. Technical knowledge

According to the features of commonly technology used in the TM development, the technology of TM is divided into three parts: composition and compatibility, production and processing Chinese crude drugs, Chinese pharmaceutical engineering.

Composition and synergy, usually refers to a mixture of Chinese Materia medica which comprises more than two Chinese herbs according to certain compatible composition rules. Composition and synergy including classical prescription[2], experiential effective recipe[3], traditional Chinese patent medicines and simple prepara-

---

[1] Guojun Xu, 2nd edition of "pharmacognosy", People's Health Publishing House, June 1997, p. 1.

[2] Classical prescription refers to the prescription which are recorded by the Eastern Han doctor named Zhangzhongjing on his book "Treatise on Cold Pathogenic and Miscellaneous Diseases".

[3] Experiential effective recipes refer to the existing prescription which is proven by clinical that the efficacy is clear.

tions[1], simple recipe, single herb, compound recipe ingredients, etc. Chinese medicine prescriptions are always organic combinations based on clinical need to use natural substances to correct the imbalance in the functions of the human body. TCM composition and synergy are the most characteristic part in the TCM knowledge, and are also the most commercially valuable part, which need to be protected affectively. It is rather difficult to protect the intellectual property of classical prescription and experiential effective recipe in fact.

The Patent Law of the People's Republic of China, which was implemented on April 1st, 1985, is the first patent law in China. At that time, the existing patent laws were only protecting methods of manufacturing drugs, not the medicines themselves. Therefore, during that period, Chinese Patent Law could not protect Chinese medicine[2].

The "People's Republic of China Patent Law" was amended for the first time on September 4th, 1992. The scope of pharmaceutical patent protection was extended from patent methods to medicine and usage patents. That is when Chinese medicine

---

〔1〕 Traditional Chinese patent medicine and simple preparations refers to the recipe which has exact efficacy, using Chinese crude drugs as raw materials, according to the TM theory, making up as a ready – made medicine by a certain dosage with regulation prescribed and standards. It can be used for doctors to cure diseases, and also can be purchased for patients have some medical knowledge. Therefore, it is widely spread to almost all of the family. It always has a specific name, appropriate packaging, indicating the efficacy, indications, usage, dosage and contraindications and precautions. There are certain quality standards and test methods, and its production must be approved by the Pharmaceutical Affairs Department. Almost come from TCM classics, as well as some from the experience side and the development side. Also, it has the advantages of no boiling, easy to carry, store, use, such as can be large – scale production.

〔2〕 Article 25. No patent right shall be granted for any of the following: ①scientific discoveries; ②rules and methods for mental activities; ③methods for the diagnosis or treatment of diseases; ④foods, beverages and condiments; ⑤pharmaceutical products, and substances obtained by means of a chemical process; ⑥animal and plant varieties; and⑦substances obtained by means of nuclear fission.

For the processes used in the manufacturing of the products listed in items④to⑥of the preceding paragraph, a patent right may be granted in accordance with the provisions of this Law.

finally became the object of patent protection[1].

Techniques for production and processing Chinese crude drugs mainly include cultivating, farming, harvesting, processing, conserving, and other technologies involved in the process of Chinese crude drugs. Concocting refers to the necessary processing procedures before the material is applied or made into various forms. It includes general pruning and organizing of raw herbs and special treatments for some materials. Preparing and processing Chinese crude drugs are traditional forms of processing, which can change the traits, taste, and effect of the medicine. Production and processing Chinese crude drugs mainly include protecting authentic ingredients (distribution and reserves of resources), techniques of Chinese materia medica cultivation, farming and production, techniques of Chinese crude drugs packaging and warehousing, the traditional processing methods, characterization, and new medicinal parts and using, the new Pieces processing, preservation technology.

Chinese pharmaceutical engineering includes pharmaceutical technology, pharmaceutical machinery and equipment, preparation of materials, automation technology, the utilization of drugs and pollution treatment technology, etc.

Obviously, the scope of patent protection in TM should include composition and synergy, production and processing Chinese crude drugs, Chinese pharmaceutical engineering. Here we will study the existing patent system problems of the protection of Chinese medicine from a practical perspective.

Section two: The Problem of Patent protection for TCM knowledge

Patent protection is one of the most effective protection ways against chemicals, but for the protection of TM is not satisfactory. According to Article 22 of the People's Republic of China Patent Law, patentable inventions must possess novelty,

---

[1] Article 25. For any of the following, no patent right shall be granted: ①scientific discoveries; ②rules and methods for mental activities; ③methods for the diagnosis or for the treatment of diseases; ④animal and plant varieties; ⑤substances obtained by means of nuclear transformation.

For processes used in producing products referred to in item④of the preceding paragraph, patent right may be granted in accordance with the provisions of this Law.

creativity and practicality. Now we will discuss the problems for patent protection of TCM knowledge according to the following three properties.

1. Novelty

1. 1. Recognized standards

China's first patent law which was implemented on April 1st, 1985, adopted "double novelty standards". According to paragraph 2 of Article 22, novelty means that before the date of filing for patent protection there's no identical invention existing domestically or abroad, or is not publicly used or otherwise known to the public, or no application has been submitted to the patent office for the same invention made by others, and should be recorded on the patent application documents after the date of filling. That is, if it is publicly disclosed, it will be based on the prior date when it was made public, either worldwide or domestically, prior to the filing date [1].

The "People's Republic of China Patent Law" 3rd amendment on Dec 27, 2008, adopted "the absolutely novelty standard". According to paragraph 2 of Article 22, novelty means that the invention does not belong to previous technology, nor does it either have any units or individual applications filed for patent at the administrative department of the State Council. And the application date should be recorded after being filed. That means before the patent filling date in any public form (publicly disclosed, publicity and other ways to be made to the public) anywhere in the world constitutes prior art. This recognized standard is similar to French patent law [2].

In practice, the national patent offices follow the identity reviewing principle to

[1] Passed on the March 12th in 1984 on the fourth meeting of the Standing Committee of the Sixth National People' Congress, starting implementation on the April 1st in 1985.

[2] Passed on the March 12th in 1984 on the fourth meeting of the Standing Committee of the Sixth National People' Congress, be amend for the first time on the September 4th in 1992 on the National People's Congress Standing, the second amendment on the August 25th in 2000 on the National People's Congress Standing, the third amendment on the December 27th in 2008 on the National People's Congress Standing.

go through novelty examination of Chinese Materia medica for patent applications[1]. In other words, only when an identical technical solution was recorded in the prior art before the application date, will the application lose its novelty.

1. 2. Existing problems

TCM already has over several thousand years of history. Much technical TCM knowledge, such as processing technology, cultivation techniques, farming techniques, pieces processing technology of Chinese Materia medica, has all been documented in ancient Chinese medicine and modern classics medicine texts. With regards to the classical prescription, a lot of texts have been diclosed to the public. Some have been developed as drugs sold to the market or used in clinics, and there are even many companies producing and selling Chinese Materia medica products. According to the novelty requirements in the patent law, this Chinese medicine technical knowledge has no novelty[2].

In this way, the most commonly used classical prescription in Chinese medicine is directly excluded from the scope of protection of the patent system. The classical prescription is the essence of TCM for a long - term to prevent and treat diseases, which has experienced repeated practice clinical for ages, groups, parties precises, and has a significant effect[3]. Many of them are the ancestral agent of ascendants compound. Since the recipe of the classical prescription has been documented in the medical books, some even spread overseas. According to the current review guidelines for the principles of novelty, all of the classical prescription lost novelty, una-

[1] Guidelines for patent examination, satte intellectual property office of the people's republic of China, 2010, p. 176. "Where the claimed invention or utility model is completely indentical with the technical contents disclosed in a reference document, or there are only simple changes in wording between them, the invention or utility model does not passess novelty. Furthermore, the meaning of indentical contents shall be construed as including the technical content directly and unambiguously derivable from the reference document."

[2] Handong Wu, *Evaluation and Legislative Proposals of Intellectual Property System of the People's Republic of China*, Intellectual Property Press, 2008, p. 339.

[3] Jinxian Gu, *Intellectual Property Protection of TM*, Tianjin People's Publishing House, 2007, p. 138.

ble to obtain patent protection, anyone can free use, without indicating the source.

Meanwhile, TCM pay great attention to the clinical practice, the efficacy of prescription be ensured only after many patient clinical validation, but due to the current review standard, prescribing behavior is "public use" the became technology of public knowledge, and therefore loss the novelty, can not be patented.

There is another question to investigate. If one technology of Chinese Materia medica is disclosed within only local or indigenous communities, should this technology be considered as "public" in the public domain, or as "non - public" and a novelty? On this issue, there may be legal uncertainties.

Chinese materia medica is based on the TCM theory, its medical uses express using the term of TCM. If there is a new requirement to protect by Western medicine illness, how to use the prior art to be compared with the terms of TCM, how to determine its novelty, there is no evaluation criteria be given in the review guidelines.

The particularity of the Chinese materia medica but also in its use position, different use parts may have different medical effects, also may have the same effect with different intensity. If the indication of the patent application of Chinese materia medica are identical with the prior art patent, the difference is only in the use part of the medicinal plants, should be how to determine its novelty, there is also no relevant regulations in the current patent examination guidelines.

2. Creativity

2. 1. Recognized standards

According to paragraph 3, Article 22, "creativity" means that in comparison with previous inventions, the new invention has prominent substantive features and notable progress. Only when an invention has simultaneously satisfied both conditions can it be classified as containing creative features[1].

---

[1] "People's Republic of China Patent Law", December 27th in 2008 the Standing Committee of National People's Congress amended version, Article 22, paragraph 3.

That an invention has prominent substantive featrues means that, having regard to the prior art, it is non – obvious to a person skilled in the art. If the person skilled in the art can obtain the invention just by logical analysis, inference, or limited exprimentation on the basis of the prior art, the invention is obvious and therefore has no prominent sabstantive features[1].

That an invention represents notable progress means that the invention can produce advantageous technical effect as compared with the prior art[2]. For example, the new invention might overcome some shortcomings and deficiencies of previous inventions, or provide new ideas or technical solutions to problems, or with regards to new technology trends.

In fact, patent examination for creativeness among Chinese Materia medica usually refers to chemical standards, the link between prescriptions with different traditional Chinese Meteria medica according to TCM theory is not to be considered.

2. 2. Existing problems

Chinese Meteria medica compound is composed of a number of Chinese crude drugs with each one having different roles, but working together comprehensively. The invention of Chinese Meteria medical compounds involves mostly either the change of ingredients or the amounts. Due to the relatively unknown chemical complexity of Chinese Meteria medica, unknown chemical composition, the large number, once you increase or decrease one or a few or change their dose, the clinical applications will change considerably. During the creativity review, the existing review standard requires the applicant to provide proof of the date when experimental efficacy was performed. This applies to experiments verifying the therapeutic effect of Chinese Material medica during clinical applications, rather than from animal experi-

---

[1] Guidelines for patent examination, satte intellectual property office of the People's Republic of China, 2010, p. 193.

[2] Guidelines for patent examination, satte intellectual property office of the People's Republic of China, 2010, pp. 193 – 194.

ments. Therefore, the result is that there is no material proving that the product brings significant creativeness.

In addition, most of the Chinese Materia medica compounds are formed on the basis of classical prescription. They involve either increasing or decreasing herbs or quantities in order to produce new prescription medicine. The Chinese Meteria medica compounds produced in this way, as compared with the previous Chinese Meteria medica technology, will have difficulty meeting the creativeness requirement. Even if it is eventually approved, since this Chinese Materia medica compound has low creativeness according to the standard of the existing patent law, it can only be classified under A61K35/00 which refers to a large group of pharmaceutical formulations which contain raw materials, or for which the structure responsible for the reaction product is unknown therefore obtaining weak patent protection.

In practice, the inspector do creative reviewing the patent application, especially the compound patent applications. Of course, thought that there was no similarities between the patent application in each group party or each group of parties and ancient Chinese medicine, creative, patent may be granted. Such review criteria leads to a lot of Chinese materia medica after it patent, others on the basis to be simple demolition party or group of parties has applied for a patent and received authorization, resulting in a large number of "similar recipes." This greatly affected the initiative of TCM knowledge holders to apply for patent protection. According to TCM theory system, this "similar recipes" usually can be derived by the technical staffs who master the TM theory, which does not meet the required creative "non - obviousness" standard.

What we should pay more attention to is examining why TCM knowledge cannot get patent protection, meanwhile "foreign Chinese Materia medica" overseas is making use of the patent system to take up TCM knowledge. These "foreign Chinese Materia medica" has two following situations:

(1) Patent protecting for the chemical structures and extraction method of ex-

tracting effective monomers or the effective parst of herbs, such as the famous case of Artemisinin[1]. Artemisia annua is a weed growing everywhere around the China. The earliest recorded method of treating malaria with Artesunate in the Eastern Jin Dynasty's "elbow emergencies". Chinese scholars began studying the anti - malarial drugs in 1969, and continued until 1972. They isolated an anti - malarial additive from the Artemisia annua, which was named Artemisinin. However, due to their lack of basic awareness of intellectual property rights, they did not apply for Artermisinin's patent protection. Consequently, foreign - based companies applied for a series of patent applications regarding the invention of Artemisinin - related derivatives, chemical synthesis and compounds, eventually making the country suffer a great loss.

(2) Patent protection for the change in dosage of original compound medicine

The formulation of the traditional Chinese herbal medicine is based on boiling a decoction, while modern Chinese medicine is made in easier forms such as tablets, capsules, granules, oral medication, injections, etc. However, modern Chinese medicine also has some shortcomings. With regard to its ingredients, the dosage ratio is fixed and cannot be as flexible as in a decoction of herbs. In recent years, there have been increasing reports about toxic and allergic reactions caused by Chinese medicine. For example, cinnabar sedative pills causing stomatitis, proteinuria and severe drug - induced colitis. Prolonged use of black tin Dan causing severe lead poisoning and oral administration of Niuhuangjiedu pill causing allergic thrombocytopenia, atopic cystitis and allergic dermatitis. Furthermore, it has been reported that taking Lingqiaojiedu pills or Yinqiaojiedu pills can cause severe anaphylactic shock.

Both methods discussed above took advantage of TCM knowledge to obtain patents, but did not have any incentive to promote the development of TCM. Contrari-

[1] Xiaoling Tian, "Study of the Chinese Compound Prescription Patent Protection", *Intellectual Property Research (Southwest University for Nationalities)*, 2007 (186): p. 122.

ly, they denied the TCM features to some extent, with distinctive "non - TCM" characteristics.

3. Practicality

3. 1. Recognized standards

According to paragraph 4, Article 22 of "the People's Republic of China Patent Law", "practicality" means that the invention is made, used and produces positive results. The method for patenting Chinese Materia medica, including its products and medical effects, is that the drug should have precise and stable medical efficacy, and repeatedly achieves its therapeutic purposes[1].

During the practicality review, the technology program for requiring the applicant of Chinese materia medica can be repeated for the application is used to achieve the purpose. And this repetition does not rely on the various random factors, and the results should be the same or similar.

3. 2 Existing Problems

Most Chinese Materia medica compounds come from plants and animals. Some are for special medical purposes and often use special materials, such as rare animals and plants, etc. Those materials are often difficult to obtain, and thus unless we use advanced science and technology to find alternatives to use instead, the actual operational use of these materials for industrial mass production is almost zero.

For example, one Chinese Materia medica compound medicine has been using osmanthus wine brewed for centuries as one of its raw materials. This raw material takes hundreds of years to be prepared. If we cannot find an alternative to the material by using advanced science and technology, the Chinese Materia medica compound with this material will have no practical use as it will not be reproducible industrially speaking.

Moreover, although there are nearly one thousand years of the practice and

---

[1] Mingde Li, Ying Du, *Intellectual Property Law*, Law Press, 2007, p. 139.

techniques for preparing and processing traditional Chinese crude drugs, many one still in private workshop stage, such formulations and production methods cannot meet the requirements of current review standard for practicality because of the manual operation of personalized. That is a doctor's prescription in accordance with the requirements of novelty and inventiveness, and the innovation generated by this personalized medicine which often do not have general applicability and cannot be created and used in the industrialization[1]. It is difficult to repeat reproduction[2].

Upon the requirements of three above points lead to, TM are often excluded from patent protection, and because of the characteristics of their own medicine leads to find patent infringement very difficult.

Western medicine all have specific chemical structural formula, when the chemicals patented can be covered by the results derivatives, if there's imitation can easily be found. The Chinese materia medica must clear the whole prescription, composition, metering and so on. And changing the prescription, such as addition or subtraction the prescription or the dose can become a new drug. That brought great difficulties for TM patent protection.

Chinese materia medica, during its production, several kinds of material mixed together, these substances may occur complex chemical reactions during processing. The taste of the Chinese materia medica is not only relevant with the recipe and production process, but also with the raw produce material. For example, after preparing of the Chinese medicine neither tablet nor decoction, even use the most advanced instruments cannot analyse its original formulation and production process. Therefore, in practice, the rights of people think that others might infringe their patents, but cannot be compared others the technical features of the product with their own,

---

[1] Xiaoting Song, *Intellectual Property Protection of Traditional Chinese Medicine Guide*, Intellectual Property Press, 2008, p. 109.

[2] Qingkui Zhang, *Compose and Review of Medical and Biological Fields Invention Patent Application Documents*, Intellectual Property Press, 2002, p. 318.

and therefore unable to prove the infringement of others. Even if they analysis out that others drugs contain several identical compounds with their medicines, but blindly Chinese materia medica often contain hundreds or even thousands of compounds and one compound can be obtained from different channels or other formulations. Therefore behavior of others cannot prove invasion[1]. Due to the particularity of TM, in practice, the phenomenon that right people cannot be protected occurred frequently.

Overall, there is a fundamental conflict between the patent system that originated in the West and TCM knowledge originating in the East. This leads to an unsatisfying result in protecting patents in respect of protecting TCM knowledge. How do we better protect TCM knowledge? Which mechanisms of protection can really play a stronger role in protecting and encouraging of TCM knowledge? These not only improve the existing "Patent Law" and "patent examination guidelines" for review Chinese materia medica, should also be comprehensive analysis the particularity of Chinese materia medica knowledge, make a special set of protection regime for TM.

---

[1] Xianbin Yang, Renwei Qian, Zhe Su, "Chinese Herbal Formula Patented Creativity Finds", Journal of Yangzhou College of Education, Mar 2009, vol 27, p. 45.

A COMPARATIVE STUDY OF MEDICAL MALPRACTICE LIABILITY BETWEEN CHINA AND FRANCE

MAN Hongjie [1]

Abstract:

Medical malpractice liability is a relatively new but rapid growing area of law since the 20th century both in France and in China. Tort Law of China, in 2009, also reaffirmed that medical malpractice liability is a type of fault - based tort liability. It adapted an objective standard of fault, as well as making provisions of presumption of fault under certain circumstance. The new law also overthrew the regulation of the Supreme People's Court, which required the defendant to prove lack of causation. In malpractice of medical products and blood transfusion, the physician will assume no - fault liability. The informed consent doctrine was introduce to China in the early 1990s and reaffirmed in the recently promulgated Tort Law. However, it has to face the conflicts of tradition and culture between China and the West in the way of making medical decisions, and also the reality of limited medical access to the large population in China.

In France, the law of medical malpractice differs according to the health provider's pubic or private nature. In civil court, dealing with action against private providers, contractual liability applies. French courts use "loss of chance" theory in deciding causation. In 2002, a no - fault compensation scheme, for patients of serious loss, was established by law.

Key words: Medical Malpractice, Liability, Comparative Study, China, France

---

[1] Associate Professor, Shandong University Law School.

Introduction

Medical malpractice liability is a relatively new phenomenon in the long history of legal liability. In Europe and the United States, it was not until the 20th century that the first cases were reported. In China, the history of medical malpractice liability can only be traced back to the 1990s. After World War II, the amount of medical malpractice cases has increased very rapidly in many countries. This reflects the development in medical and biological science accompanied by the achievement of the patients' right movement, which is part of the civil rights tide. Both China and France have a civil law tradition, which takes medical malpractice as a category of civil liability. As a result of comparative research, this chapter will review the legislation and legal practices in China and France.

1. Medical Malpractice Liability in China

1. 1. Overview

The Chinese law, regards of medical malpractice liability, is also a fault - based tort. Article 54 of the Tort Law says: "Where a patient sustains any harm during diagnosis and treatment, if the medical institution or any of its medical staff is at fault, the medical institution shall assume the compensatory liability." And Article 6 says: "One who is at fault for infringement upon a civil right or interest of another person shall be subject to the tort liability." Medical malpractice liability is also based on the fault of physicians. In the Chinese health care system, almost all the physicians practice in hospitals or other types of institutions. In accordance with Article 54, it is the institution, in stead of the physician himself, who is responsible for the physician's malpractice.

1. 2. Fault

As in other civil law countries, fault is regarded as a subjective matter. Traditionally, fault is described as a mental activity, either being too careless to notice the risk—namely negligence—or, being too confident to avoid the damage—namely overconfidence. In medical malpractice, courts found that it is usually too difficult for the patient to prove the physician's fault merely, from the subjective point of view. Therefore, the Tort Law introduced "Objective" theory of fault. Article 57 says, "Where any medical staff member fails to fulfill the obligations of diagnosis and treatment up to the standard at the time of the diagnosis and treatment and causes any harm to a patient, the medical institution shall assume the compensatory liability." In order to judge whether the physician is at fault, an objective standard was developed. This standard requires the physician to perform his diagnosis or treatment according to the normal, local level of medical professionals, at the time of performance.

Article 58 is also the facto per se clause of fault. It says, "under any of the following circumstances, a medical institution shall be at fault for any harm caused to a patient: 1. violating a law, administrative regulation or rule, or any other provision on the procedures and standards for diagnosis and treatment; 2. concealing or refusing to provide the medical history data related to a dispute; or, 3. forging, tampering or destroying any medical history data." Therefore, when any of these situations occurs, the physician's fault is assumed automatically. The patient is exempted from the duty of proving the physician's fault in accordance with Article 57. There is a debate, among the legal scholars, about whether this assumption of fault is disprovable by the physician. The majority of scholars agreed that this assumption can not be disproved, leaving no space for the physician to deny his fault under such circumstances.

1.3. Causation

Similar to other civil law countries, causation is the legal and logical connection between the infringer's wrongful behavior and the injury suffered by the infringed. In both a fault-based liability or in strict liability, it is the plaintiff's duty to prove the causation. In medical malpractice cases, the causation is always difficult to prove and it is not easy to tell if the damage caused the patient is the result of the physician's malpractice, or it is merely the natural consequence of patient's disease or injury. It is even more difficult for the patient who is usually a layman without any knowledge or training of medical science to do so. The courts tend to shift the burden of proof for causation onto the physician. In 2001, the People's Supreme Court promulgated the Relations Concerning Evidence in Civil Procedure. Para 8 Article 4 of the regulation reads, "In the tort liability suit caused by medical malpractice, it is the duty of medical institution to prove that there was not causation between the malpractice and the damage of the patients, as well as there is no fault in the health care provided." This paragraph has generated numerous subsequent debates. Physicians argued that this regulation was far too severe to them, for the causation was unclear in many cases, even from the professional point of view. They should not assume all the risks of uncertainties in health care. Otherwise a medical career would be unreasonably risky. It turned out that this regulation, of burden of proof, was overruled by the Tort Law in 2009. By reconfirming medical malpractice liability is a fault – based liability, Article 54 returned the traditional orbit of distribution of burden of proof. It is the patient's duty to prove the causation according to Tort Law. Discussion regarding balancing the unequal ability of proof, between the patient and physician, is continuing.

1.4. Strict Liability

In the Chinese law, there is also space for strict liability in medical malpractice. Article 59 of the Tort Law says, "Where any harm to a patient is caused by the defect of any drug, medical disinfectant or medical instrument or by the transfusion of substandard blood, the patient may require compensation from the manufacturer or institution providing blood, or require compensation from the medical institution. If the patient requires compensation from the medical institution, the medical institution that has paid the compensation shall be entitled to be reimbursed by the liable manufacturer or institution providing blood." In case of medical injury, caused by medical product defects or blood transfusion, the medical institution will assume strict liability to the patient, even if it successfully proved there is no fault by the physician during the health care service.

1.5. Compensation

In accordance with Article 16 of Tort Law, "Where a tort causes any personal injury to another person, the tortfeasor shall compensate the victim for the reasonable costs and expenses for treatment and rehabilitation, such as medical treatment expenses, nursing fees and travel expenses, as well as the lost wages. If the victim suffers any disability, the tortfeasor shall also pay the costs of disability assistance equipment for the living of the victim and the disability indemnity. If it causes the death of the victim, the tortfeasor shall also pay the funeral service fees and the death compensation." Thus, the scope of compensation for medical malpractice includes: ①Costs for health care, including treatment and nursing cost; ②Loss of income, including wages loss and loss of potential working ability in case of disability; and, ③Death compensation when the victim dies.

Tort Law also allows for immaterial compensation. Article 22 says "Where any harm caused by a tort to a personal right or interest of another person inflicts a serious mental distress on the victim of the tort, the victim of the tort may require compensation for the infliction of mental distress." In case that mental distress, caused by the medical malpractice, is serious, the patient is entitled the right to claim immaterial compensation as well.

2. Informed Consent and Tort Liability in China

2.1. Legal Resource

Informed consent is the dominant doctrine controlling the physician - patient relationship in the modern world. Chinese history of informed consent can only be traced back to the Regulation Governing the Administration of Medical Institutes by the State Council in 1994. Article 33 of the Regulation says, "Any operation, invasive diagnosis or treatment suggested by the medical institute should get the consent of the patient, as well as the signature of patient's family member or the next to kin; in case the consent of the patient is not available, the signature of the patient's family member or the next of kin is necessary; in case neither consent nor signature is available, or in a special emergency, the doctor should prepare a proposal and carry it out only after the proposal is approved by the head of the medical institute or other authorized staff." It was the first time that a Chinese decree recognized the duty of obtaining consent from the patient, or his family, before treatment of diagnosis. However, no duty of informing is required.

In 1998, the Law on Licensed Medical Practitioners passed by the Standing Committee of National People's Congress reaffirmed informed consent. Article 26 says "the physician should introduce the patient's condition to the patient or his family, while avoiding harmful consequence to the patient." According to this statute, the physician has a duty of "introducing" rather than "informing".

Article 55 of Tort Law states, "During the diagnosis and the treatment, the medical staff shall explain the illness condition and the relevant medical measure to the patient. If any operation, special examination or special treatment is needed, the medical staff shall explain the medical risks, alternate medical plan and other information to the patient in a timely manner, and obtain a written consent from the patient. In case it is not suitable to inform the patient, the patient's family member shall be informed and give the written consent instead. When a medical staff breaches the duty above, the medical institute shall be found liable for any patient's damage caused by it." Article 56 says, "In case of emergency, when the written consent from the patient or his family member is not available, necessary treatment shall be taken only if the proposed medical plan is approved by the head of the medical institute or an authorized person."

2.2. The Chinese Reality of Informed Consent

Although the informed consent doctrine was rapidly introduced to China and adopted by the Chinese statutes in the past two decades, its effect is still somewhat doubtful.

2.2.1. The Cultural Conflict and impact

There is a cultural conflict between China and the West, in the way of making medical decisions. The informed consent doctrine is firmly rooted in western culture and tradition. The belief and spirit of autonomy, which is the fundamental ethical basis of informed consent, is derived from Jewish – Christian culture and a philosophy concentrating on individualism. Comparatively, Chinese culture is much more of collectivism. In ancient Chinese (probably even of East Asia) tradition, families, rather than the individual person, were the most important units in the society. Although things changed a lot in the past century, families still play a key role in many ways, e. g. in the area of medical issue. In China, it is more common to leave the right to know and decide in medical issues to the family numbers, rather than the patient. Usually, it is not the patient but his family who is informed of the patient's medical condition and the benefits, risks, alternatives and other information concerning the proposed treatment. It is also the patient's family who makes the decision and authorizes doctors to proceed in the treatment thereafter. This tradition justifies the earlier statutes' valuing the family's consent and signature more than that of the patient. In Chinese medical practice, it is probably fine for the physician to give the patient the information in case it is only a cold or high blood pressure. Nevertheless, if it is cancer, or other severe or mortal illness, the physician will never tell the patient himself. The physician may even complain about the by the patient's family if he tells the patient the truth. Commonly, it is the patient's family who make the final decision. Sometimes it drives the physician into a dilemma to choose to obey the law or respect the tradition.

What makes things even worse, is the situation when the family member is not able to make a rational choice for the patient, or even may have a conflict of interest with the patient. For most cases, the family will do the best for the patients and make a favorable decision. Sometimes the family can not rationally decide, which may do harm to the patient. At the end of 2007, there was a well – known case in China called "Xiao Zhijun event". Xiao Zhijun was a low – income peasant from

South China, living in Beijing. His seven – month – pregnant wife caught pneumonia and fell into a coma. After Xiao delivered her to a hospital, the physicians urged that a caesarean be performed without any delay, to save the life of both the mother and the child. Nevertheless, Xiao did not believe what he was told and refused the operation, in spite of the physicians' repeated persuasion. Neither the mother nor the child survived. It caused tremendous argument about the justification and rationality of the family's role in making a medical decision. It can be more problematic when there is conflict of interests between the family member and the patient.

2. 2. 2. How is informed consent regarded?

On the other hand, does the Chinese patient really understand the issues about which they are informed? Because of the strained relationship between patients and physicians, and the heavy burden of legal liability for the medical industry, physicians tend to treat the informed – consent as a defense of their professional risks or as an excuse in malpractice law suits, rather than the mechanism of respecting patients' autonomy or protecting patients' rights. Therefore, the informed consent is simplified as a pile of forms and clauses which are far beyond patients' understanding and knowledge. In this case, informed consent does not contribute to the construction of a trusting relationship between doctors and patients; on the contrary it makes them feel estranged.

2. 2. 3. Informed Consent Facing the Limited Medical Access

There is another thing that may have a negative effect on the practice of informed consent, the limited medical access in China. With the largest population on the globe, China is still a source – limited society concerning health service. According to World Health Statistics 2010 by the WHO, the numbers of physicians, nursing personnel, dentistry personnel and pharmaceutical personnel per 10, 000 population in China are 14, 10, 1 and 3, respectively while the American numbers are 27, 98, 16 and 9. There are too many patients for the physician to treat which causes the physician to lose patience. When a physician hardly has his time to examine a patient and give a prescription, how can one expect that physician to spare his precious time to tell the patient everything he found and wait for the patient's decision before further diagnosis or treatment is performed? Furthermore, unlike in the U. S. , Chinese people do not have a private physician of their own, which means it is much more difficult to establish a trusting relationship between the patient and the physician (not "his physician"), when they are utterly strange to each other.

3. Medical Malpractice Liability in France

3. 1. Overview

French healthcare service is divided into two categories, pubic providers and private providers, who are governed by separate jurisdictions. When the patient is treated by a private provider, the action for compensation is brought in a civil court while an action against public providers should be brought in an administrative law court.

The legal basis of the relationship between the patient and a private health provider is governed by contract. It is considered that there is a contract for health service between the patient and the provider. An action against the health provider is based on breach of contract. This doesn't mean that the provider has a duty to cure the patient. As the court stated Mercier in 1936, the contract imposes a duty on the provider "not to cure the patient, but to provide him in a conscientious and attentive way with the appropriate care in accordance with the current state of scientific knowledge."[1] The provider will be found liable for the patient's damage only if he fails to provide reasonable care and skill. This is based on Article 1147 of the French Civil Code. If emergency health care was provided to a patient, or the family of the patient sues the provider on their own behalf, there will be no contract between the two parties, in which case the action will base on tort. This type of fault – based tort

[1] Case civ. 20 May, 1936, Mercier, D. 1936. 1. 88, concl. Matter, rapp. Josserand. See Ewoud Hondius, eds. , *The Development of Medical Liability*, Cambridge: Cambridge University Press, 2010, p. 80.

is governed by Article 1382 of the French Civil Code. It does not make too much difference that the action is based on contract or on tort, concerning the outcome.

The action against a public provider is considered a lawsuit against public authority. There is no contact between the patient and the public providers. The case will be tried in accordance with administrative principles. Before 2002, there was little statutory law provision concerning medical malpractice liability. Most of the regulations were established by courts, either by the civil court or by the administrative court.

In 2002, there was a dramatic change of legislation. On 4 March of that year, "the Act No. 2002 – 303 concerning the rights of patients and the quality of the health system" was approved by the French Parliament. Later in the same year, the law of 31 December 2002 made subsequent amendment to the Code de la santé publique (Code of Public Health). This law introduced a new legal framework of medical malpractice compensation which was no longer based on fault, and unified the liability for public and private providers. It only applied when the patient suffered serious accidents, therefore the traditional court – made regulations remained valid.

3. 2. Fault

In traditional jurisdiction, irrespectively whether in civil law or in administrative law, medical malpractice liability is fault – based. In 1835, Cour de cassation (the supreme court of France) found a doctor liable in tort, under Article 1382 and 1383 of the Civil Code, for grave negligence.[1] In this case, grave negligence was defined as "carelessness or not knowing of things it is obvious he should be aware of." The court used the term "grave negligence" to express an attitude of being unwilling to interfere with medical science. Only when the negligence is so obvious that there was no need to refer to medical professional knowledge, the defendant will be found liable. In the Mercier, the court stated that the doctor's failure to provide reasonable

[1] Cass. Req. 16 June 1835, S. 1835. 1. 402. See supra 1, p. 78.

care and skill is a breach of contract. Contract liability is also based on fault. In the administrative law system, the public hospital would be liable only if a faute lourde was found. The term "Faute lourde", used by administrative courts, also means serious fault. In the early cases, both the civil and the administrative courts were very strict with recognition of the health provider's fault, due to a medical paternalism view of doctor/patient relationship. In the recent half century, both the Cour de cassation and the Cousel d'Etat[1] showed more interests in protecting patents' right, which made it easier to establish the provider's fault. The doctor's behavior, in diagnosis and treatment, is assessed by the court, in accordance with the prevailing level of medical scientific knowledge at the time the damage occurred. Usually, this test will rely on evidence offered by an expert witness appointed by the court. In administrative cases, the court uses a "reasonable practitioner" standard to define the doctor's fault. Failure to obtain consent from the patient, before a medical act was performed, will constitute a fault, according to Article 16 – 3 of the Civil Code.

3. 3. Causation

It is the patient's duty, in a medical malpractice action, to prove the causal link between the doctor's wrongful behavior and the patient's damage. From 1970s, the civil courts, as well as the administrative courts introduced the term and technique of "loss of chance" concerning causation. When the patient is unable to prove causation, and the court finds that the patient would have had a great chance to survive or recover were it not for the doctor's fault, the court will allow the patient to get compensation of certain percentage of total loss, according to its assessment of the percentage chance of the patient's survival or recovery.

3. 4. Strict Liability

Although the whole system of medical malpractice liability is based on fault,

---

[1] Cousel d'Etat is a body of the French national government that provides the executive branch with legal advice and acts as the administrative court of last resort.

there are two exceptions where strict liability applies. One is hospital infection liability, when the damage to the patient was caused by a bacterial infection, during his stay in a hospital or other medical facilities, the hospital or other health provider will be found liable without considering any fault in the practice undertaken. The only exemption to liability is to prove that the infection was caused outside the hospital or other medical facilities. The other exception is liability for defective medical products. Liability for products is governed by the Product Liability Directive (85/374/EEC) of EU and French Product Liability Act, for which both provide strict liability.

3.5. No - fault Compensation Scheme by the Law of 4 March 2002

The Law of 4 March 2002 is a result of long time proposed reform in medical law. Part IV "Indemnification of the consequences of sanitary risks" (Article 98 - 107) set up a scheme which guarantees compensation to victims of serious medical accidents. According to this scheme, the patient may get compensation for their loss from a medical accident only if their injury is sufficiently serious. In order to tell how serious the injury is, there is a predefined level of injury, which is decided by the patient's level of permanent disability or the duration of temporary inability to work. A percentage system is used to illustrate this disability level, and currently requires at least 24% of disability, which means the damage caused an inability to work of more than 6 months, or particularly serious disorders in the patient's life conditions or prevented him from resuming his professional activity. When the minimum level of seriousness is satisfied, the patient may seek compensation through this scheme. Otherwise, he has to sue the health provider, in a civil or administrative court, for malpractice liability.

The Law of 4 March, 2002 created several authorities for this purpose. The National Compensation Office for Medical Accidents, Iatrogenic Diseases and Nosocomial Infections (ONIAM) is the national level authority in charge of compensation for medical accidents, as well as its funding. It is also the ONIAM's duty to supervise Regional Commissions for Conciliation and Compensation (CRCIs). CRCIs are the organizations established, on a regional basis, to deal with claims for compensations. A CRCI, chaired by a judge, consists of 20 members, including representatives for patients, medical professionals, administrators, the national compensation fund and insurers.

When a medical accident occurs, the patient may file a claim to a CRCI. If a fault by the health provider (in prevention, diagnosis, treatment or other professional behavior) is found, the CRCI will issue an opinion that the provider should be liable for the patient's damage. In this case, the provider's insurer (indemnity insurance for health providers is compulsory in France) must make an offer of compensation to the patient. If the insurer refuses to make an offer, or the amount of compensation exceed the coverage amount of the insurance, the patient will get the compensation from the State fund of France. Under such circumstance, the State fund takes over the right to demand the money from the insurer. In case the insurer and the patient are unable to reach a compromise regarding the amount of compensation, the patient is entitled to sue the insurer in a court.

If there is no fault found in the practice, the patient will get the compensation from the State fund. The patient is also authorized to sue the State fund if he considers the compensation from the fund is insufficient.

4. Conclusion

Generally speaking, Chinese and French systems concerning medical malpractice liability bear more commons than differences. Both countries have civil tradition and take the medical malpractice as a form of obligation, either by contract law or by tort law. In order to seek compensation from the health care provider, the patient has to build up his/her claim by proving fault, damage, malpractice and causation. However, the acceptance of loss of chance doctrine in France gives more flexibilities in the recognition of causation when the accurate causal link is not clear. Furthermore, the non – fault compensation scheme established by the Law of 4 March,

2002 guarantees the patients suffered serious damage to get compensation in a timely manner, which is important to achieve the aim of corrective justice, which is the dominant aim of the malpractice law.